Recent Advances in Cellular D2D Communications

Special Issue Editors

Boon-Chong Seet

Syed Faraz Hasan

Peter Han Joo Chong

MDPI • Basel • Beijing • Wuhan • Barcelona • Belgrade

MDPI

Special Issue Editors
Boon-Chong Seet
Auckland University of Technology
New Zealand

Syed Faraz Hasan
Massey University
New Zealand

Peter Han Joo Chong
Auckland University of Technology
New Zealand

Editorial Office
MDPI AG
St. Alban-Anlage 66
Basel, Switzerland

This edition is a reprint of the Special Issue published online in the open access journal *Future Internet* (ISSN ISSN 2072-6651) from 2017–2018 (available at: http://www.mdpi.com/journal/futureinternet/special issues/Recent Advances).

For citation purposes, cite each article independently as indicated on the article page online and as indicated below:

Lastname, F.M.; Lastname, F.M. Article title. *Journal Name*. **Year**. *Article number*, page range.

First Editon 2018

ISBN 978-3-03842-738-4 (Pbk)
ISBN 978-3-03842-737-7 (PDF)

Table of Contents

About the Special Issue Editors

Boon-Chong Seet, Associate Professor, received his PhD degree in Computer Communications Engineering from Nanyang Technological University, Singapore, in 2005. Upon graduation, he worked as a Research Fellow under the Singapore-Massachusetts Institute of Technology Alliance (SMA) Program at the National University of Singapore. Since December 2007, he is with the Department of Electrical and Electronic Engineering, Auckland University of Technology, New Zealand, where he is currently an Associate Professor and leader of a wireless engineering research group. His research ac-tivities span the fields of info-communication technologies (ICT), including a number of recent works on 5G communications. He is a Senior Member of the IEEE.

Syed Faraz Hasan, Senior Lecturer, received his PhD degree from the University of Ulster, U.K., and Bachelor degree in Electrical Engineering (with distinction) from the NED University of Engineering and Technology, Pakistan, in 2011, and 2008, respectively. He previously worked at Sungkyunkwan University, South Korea, and Korea Advanced Institute of Science of Technology (KAIST). He is currently a Senior Lecturer with the School of Engineering and Advanced Technology, Massey University, New Zealand, where he leads the Telecommunication and Network Engineering research group. His research interests include device-to-device communication, energy harvesting, and software-defined networking. He is a Senior Member of IEEE.

Peter Han Joo Chong, Professor, received his BEng degree (Hons.) from Technical University of Nova Scotia, Canada, in 1993, and his MASc and PhD degrees from the University of British Columbia, Canada, in 1996, and 2000, respectively. From 2002 to 2016, he was with the School of Electrical and Electronic Engineering, Nanyang Technological University, Singapore, as an Associate Professor (Tenured). Presently, he is with the Department of Electrical and Electronic Engineering, Auckland University of Technology, New Zealand, as a Full Professor and Head of Department. His research interests are in the areas of mobile communications systems, including radio resource management, multiple access, MANETs, multihop cellular networks, and vehicular communications networks.

future internet

MDPI

Editorial

Recent Advances on Cellular D2D Communications

Boon-Chong Seet [1],* , Syed Faraz Hasan [2],* and Peter Han-Joo Chong [1],*

[1]   Department of Electrical and Electronic Engineering, Auckland University of Technology,
     Auckland 1010, New Zealand
[2]   School of Engineering and Advanced Technology, Massey University, Palmerston North 4442, New Zealand
*   Correspondence: boon-chong.seet@aut.ac.nz (B.-C.S.); F.Hasan@massey.ac.nz (S.F.H.);
     peter.chong@aut.ac.nz (P.H.-J.C.)

Received: 15 January 2018; Accepted: 16 January 2018; Published: 17 January 2018

Device-to-device (D2D) communications have attracted a great deal of attention from researchers in recent years. It is a promising technique for offloading local traffic from cellular base stations by allowing local devices, in physical proximity, to communicate directly with each other. Furthermore, through relaying, D2D is also a promising approach to enhancing service coverage at cell edges or in black spots. Besides improving network performance and service quality, D2D can open up opportunities for new proximity-based services and applications for cellular users. However, there are many challenges to realizing the full benefits of D2D. For one, minimizing the interference between legacy cellular and D2D users operating in underlay mode is still an active research issue. With the 5th generation (5G) communication systems expected to be the main data carrier for the Internet-of-Things (IoT) paradigm, the potential role of D2D and its scalability to support massive IoT devices and their machine-centric (as opposed to human-centric) communications need to be investigated. New challenges have also arisen from new enabling technologies for D2D communications, such as non-orthogonal multiple access (NOMA) and blockchain technologies, which call for new solutions to be proposed.

This special issue aims to present a collection of exciting papers, reporting the most recent advances in cellular D2D communications. Through invited and open call submissions, a total of ten excellent articles have been accepted, following a rigorous review process that required a minimum of three reviews and at least one revision round for each paper. The list of accepted articles includes one review and nine original research articles on addressing many of the aforementioned challenges and beyond.

The first paper by *Höyhtyä, Apilo* and *Lasanen* [1] is a review article that analyzed the latest energy consumption models of 3GPP standardized LTE (long-term evolution) and WiFi interfaces, with recommendations on energy saving options for D2D communications in a set of application scenarios.

Distributed resource sharing and allocation are amongst the most important issues in cellular D2D networks. *Hong, Wang, Cai* and *Leung* [2] investigated the issue of fairness in cooperative D2D computational resource sharing, and proposed a blockchain-based credit system where user's computational task cooperation are recorded on public blockchain ledger as transactions, and their credit balance can be easily accessed from the ledger. The performance of the proposed credit system is demonstrated by incorporating it into a connectivity-aware task scheduling scheme to enforce fairness among users in the D2D network.

Radio resource is another resource type that must be efficiently managed. The next four papers explore different strategies for allocating radio resources such spectrum and transmit power for D2D communications. *Jiang, Wang, Ren* and *Xu* [3] studied the problem of spectrum resource and transmit power allocation for underlay multicast D2D communications, and presented a heuristic and low-complexity resource and power allocation scheme that aims to maximize overall energy efficiency, while satisfying the QoS (quality of service) requirements of both cellular and D2D users. Similarly, for underlay D2D communications, *Ban* [4] proposed a practical scheme with low complexity and signaling overhead for distributed radio resource management. The scheme does not require any channel feedback, and each D2D pair can transmit on its own, based on simple bitmap information broadcast by the base station and an optimal threshold value derived to maximize the average sum-rates.

To address the issue of mitigating multicell D2D underlay interference, *Katsinis, Tsiropoulou* and *Papavassiliou* [5] proposed a two-step approach, which involves solving the initial resource block allocation problem by formulating it as a bilateral symmetric interaction game, and then addressing the transmit power allocation problem by using a linear programming approach to minimize the total interference of the network. In order to respond to changing network conditions, there is a need for the resource allocation mechanism to be adaptive. The paper by *Khan, Alam, Moullec* and *Yaacoub* [6] presented a cooperative reinforcement learning algorithm for adaptive allocation of resource blocks and transmit power to D2D users in a cellular network. By efficient control of the interference level, the proposed algorithm results in improved overall system throughput, D2D throughput and fairness among D2D users.

The spectrum efficiency problem in group D2D communications is next addressed by *Anwar, Seet,* and *Li* [7] who proposed a QoS based non-orthogonal multiple access (Q-NOMA) scheme in which D2D users in a NOMA transmission are ordered according to their QoS requirements. Using stochastic geometry tools, the authors modeled the spatial relationships and interferences between the group D2D users, which led to a closed-form expression for characterizing their outage performance.

In human-centric D2D communications, the social and trust relationships between users are humanistic features that can be leveraged for enabling more secure and reliable solutions. *Militano, Orsino, Araniti* and *Iera* [8] exploited the social relationships among D2D users to model the trust level between them, and proposed a social trust-based solution for enhancing the performance of D2D-enhanced cooperative content uploading in the presence of packet dropping or corrupting malicious nodes for narrowband-IoT cellular environments. In another cooperative design, *Chiti, Fantacci* and *Pierucci* [9] considered the problem of relay-assisted cooperative multicast (one-to-many) D2D communications, and presented a relay selection scheme that considers both propagation link conditions and relay's social trust level with the constraint of minimizing end-to-end delay in an integrated social–physical network.

Besides multicast, broadcast (one-to-all) is another communication option that can be supported by D2D. This special issue concludes with a paper by *Nardini, Stea* and *Virdis* [10], who proposed a message broadcast solution appropriate for vehicular networks based on multihop D2D communications. The proposed solution allows a user to specify its target area without being constrained by cell boundaries. It relies on application-level device intelligence and standard D2D resource allocation methods of LTE-A to enable fast, reliable and resource-efficient message broadcast services.

Acknowledgments: The Guest Editors wish to thank all the contributing authors, the professional reviewers for their precious help with the review assignments, and the excellent editorial support from the Future Internet journal at every stage of the publication process of this special issue.

Author Contributions: All three authors contributed to making the editorial decisions on the submissions to this special issue. B.-C.S. prepared this editorial with some inputs from S.F.H. and P.H.-J.C.

Conflicts of Interest: The authors declare no conflict of interest.

References

1. Höyhtyä, M.; Apilo, O.; Lasanen, M. Review of Latest Advances in 3GPP Standardization: D2D Communication in 5G Systems and Its Energy Consumption Models. *Future Internet* **2018**, *10*, 3. [CrossRef]
2. Hong, Z.; Wang, Z.; Cai, W.; Leung, V.C.M. Blockchain-Empowered Fair Computational Resource Sharing System in the D2D Network. *Future Internet* **2017**, *9*, 85. [CrossRef]
3. Jiang, F.; Wang, H.; Ren, H.; Xu, S. Energy-Efficient Resource and Power Allocation for Underlay Multicast Device-to-Device Transmission. *Future Internet* **2017**, *9*, 84. [CrossRef]

4. Ban, T.-W. A Practical Resource Management Scheme for Cellular Underlaid D2D Networks. *Future Internet* **2017**, *9*, 62. [CrossRef]
5. Katsinis, G.; Tsiropoulou, E.E.; Papavassiliou, S. Multicell Interference Management in Device to Device Underlay Cellular Networks. *Future Internet* **2017**, *9*, 44. [CrossRef]
6. Khan, M.I.; Alam, M.M.; Moullec, Y.L.; Yaacoub, E. Throughput-Aware Cooperative Reinforcement Learning for Adaptive Resource Allocation in Device-to-Device Communication. *Future Internet* **2017**, *9*, 72. [CrossRef]
7. Anwar, A.; Seet, B.-C.; Li, X.J. Quality of Service based NOMA Group D2D Communications. *Future Internet* **2017**, *9*, 73. [CrossRef]
8. Militano, L.; Orsino, A.; Araniti, G.; Iera, A. NB-IoT for D2D-Enhanced Content Uploading with Social Trustworthiness in 5G Systems. *Future Internet* **2017**, *9*, 31. [CrossRef]
9. Chiti, F.; Fantacci, R.; Pierucci, L. Social-Aware Relay Selection for Cooperative Multicast Device-to-Device Communications. *Future Internet* **2017**, *9*, 92. [CrossRef]
10. Nardini, G.; Stea, G.; Virdis, A. A Fast and Reliable Broadcast Service for LTE-Advanced Exploiting Multihop Device-to-Device Transmissions. *Future Internet* **2017**, *9*, 89. [CrossRef]

future internet

MDPI

Review

Review of Latest Advances in 3GPP Standardization: D2D Communication in 5G Systems and Its Energy Consumption Models

Marko Höyhtyä *, Olli Apilo and Mika Lasanen

VTT Technical Research Centre of Finland Ltd., P.O. Box 1100, FI-90571 Oulu, Finland; olli.apilo@vtt.fi (O.A.); mika.lasanen@vtt.fi (M.L.)
* Correspondence: marko.hoyhtya@vtt.fi; Tel.: +358-40-548-9204

Received: 23 November 2017; Accepted: 20 December 2017; Published: 3 January 2018

Abstract: Device-to-device (D2D) communication is an essential part of the future fifth generation (5G) system that can be seen as a "network of networks," consisting of multiple seamlessly-integrated radio access technologies (RATs). Public safety communications, autonomous driving, socially-aware networking, and infotainment services are example use cases of D2D technology. High data rate communications and use of several active air interfaces in the described network create energy consumption challenges for both base stations and the end user devices. In this paper, we review the status of 3rd Generation Partnership Project (3GPP) standardization, which is the most important standardization body for 5G systems. We define a set of application scenarios for D2D communications in 5G networks. We use the recent models of 3GPP long term evolution (LTE) and WiFi interfaces in analyzing the power consumption from both the infrastructure and user device perspectives. The results indicate that with the latest radio interfaces, the best option for energy saving is the minimization of active interfaces and sending the data with the best possible data rate. Multiple recommendations on how to exploit the results in future networks are given.

Keywords: D2D communications; 5G systems; power efficiency

1. Introduction

Device-to-device (D2D) communications in infrastructure networks have been studied actively since the 1990s [1], due to the potential to reduce delays, increase throughput, and to improve power or energy efficiency. D2D enables cooperative services and data dissemination methods and can be used in coming 5G networks over various radio access technologies (RATs). Actively developed application areas currently include 3GPP proximity services, public safety communications, vehicle-to-everything (V2X) communications, autonomous ships, the Internet of Things (IoT) and wearables [1–9]. For instance, the number of wearable devices is predicted to grow from 325 million in 2016 to 929 million in 2021, when 7% of the devices may use in-built cellular connectivity [10]. Other devices, on the other hand, may obtain cellular access through e.g., smart phones.

An essential part of the use of D2D in the mentioned application areas is energy efficiency [11–14], which is heavily dependent on the used radio interfaces. In general, the role of WiFi and other small cell technologies is important, as 60% of mobile data was offloaded onto the fixed network through WiFi or femtocell in 2016 [10]. In addition, computing power is important, especially in short distance communication [15]. Compared to theoretical power control work, such as [16,17], one is able to estimate more accurately the resource use in a practical network if measurement-based models for air interfaces are available. The power consumption of different 3GPP long-term evolution (LTE) and WiFi interfaces has been actively measured and modelled in recent years [18–22]. Both user device and base station power consumption models are available. However, there is a lot of variation in measurement campaigns between different protocols and between different smart phone models.

Some of the differences can be explained by the new generation of air interfaces and partially the power consumption changes are due to the different use of the user devices. For example, social networking [23] generates a constant stream of traffic, causing the mobile device to frequently move between idle and connected states. Energy state transitions alone cost energy, but these transitions also cause excessive signaling overhead in (3GPP) networks. Mechanisms such as adaptive discontinuous reception (DRX), user equipment (UE) assistance, energy harvesting, and massive multiple-input multiple-output (MIMO) antenna systems at the base station side have been proposed to reduce the power consumption of LTE mobiles [24–29].

We analyzed the power consumption of user devices in D2D communications in [30] and studied the power consumption from the base station perspective in [31] using many different measurement-based LTE and WiFi models. In this paper, we extend and unify analysis of [30] and [31] and update the results with the latest power consumption models [32]. In addition, we review the status in 3GPP standardization of D2D communications, focusing especially on IoT, wearables, and V2X communications [33–38]. The analysis shows where the industry is going and deepens the discussion on energy efficiency aspects in depicted networks. We believe that quality of service (QoS) and priority management mechanisms such as network slicing [35,36] can also be used to improve the performance of D2D networks.

We will extend the state-of-the art in [11–32], summarizing the novelty of this paper as: (1) Review of the status of the 3GPP standardization, including a summary of D2D features of different releases of the standard. (2) Definition of a set of D2D application scenarios with multiple data delivery options. (3) Analysis of the power consumption of the network in the depicted scenarios using measurement-based models. The 5G system will be a multi-RAT (radio access technologies (RATs)) system that enables seamless interworking between those RATs. Unlike previous works, we will consider both end user and base station perspectives in this paper. There are no measurement-based models of new 5G interfaces available yet, but there are LTE and WiFi models that will be an essential part of the coming 5G system. Therefore, we use the latest LTE-advanced and WiFi power consumption models in the analysis.

The paper is structured as follows. Section 2 reviews the status of 3GPP standardization. The system model and the use cases for analysis are defined in Section 3. The selected measurement-based power consumption models are described in Section 4. Performance analysis models from base station and end user device perspectives are depicted in Section 5, and the results given in Sections 6 and 7 provides recommendations based on the conducted analysis. Section 8 concludes the paper.

2. Advances within 3GPP Standardization on D2D

3GPP specified the basic functionalities for D2D communications in release 12, where the main motivation was to develop a global standard for public safety communications [37]. However, the application scenario of 3GPP proximity services (ProSe) was not limited to public safety, D2D extension of conventional cellular services was also considered [38]. The basic architecture of the 3GPP ProSe is shown in Figure 1. A UE (user equipment) that wants to use ProSe must first contact the ProSe function through the logical interface named PC3 to get authorization and security parameters. After the discovery request and response message exchange via PC3 is completed, the UE can start the direct discovery process to find other UEs with ProSe cabability in their proximity using the PC5 interface. When two (or more) ProSe-enabled UEs have discovered each other, they can start direct communication over the direct link between them.

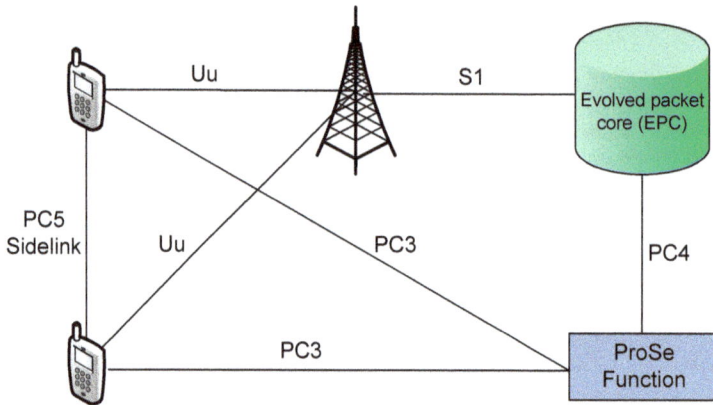

Figure 1. Architecture and logical interfaces for proximity services (ProSe); PC: Interface between ProSe components; S1: interface between base station and the core network; Uu: Air interface.

The physical interface between two ProSe UEs is called sidelink. Time-frequency resources for the sidelink are shared with the uplink (UL), and the sidelink waveform is also similar to the single-carrier frequency-division multiple access (SC-FDMA) UL waveform. As ProSe was originally designed for public safety group communications, the sidelink transmission is based on multicasting with no hybrid automatic repeat request (HARQ) feedback. Instead, each medium access control (MAC) protocol data unit (PDU) is retransmitted three times with a different redundancy version for each transmission. Dedicated resource pools are allocated for sidelink transmissions in order to avoid collisions between them and conventional UL transmissions. The subframes and physical resource blocks (PRBs) belonging to sidelink resource pools are broadcasted as system information to UEs. Resources within a resource pool can be allocated by an evolved NodeB (eNB) (Mode 1) or they can be autonomously selected by a UE (Mode 2) [39], which enables sidelink communication when a UE is not within the cell coverage. ProSe communication was further enhanced in release 13 e.g., by allowing a UE to operate as a relay for another UE. The relaying was implemented at layer three in such a simple way that the network cannot differentiate the traffic of the remote UE from that of the relay UE. This limits the ability of the operator to treat the remote UE as a separate device for billing and security [40].

Service requirements related to the 5G system [41] consider D2D in two different ways. The first one uses direct device connection without any network entity in the middle. In the second approach, a relay UE is between a UE and the 5G network. This is called indirect network connection mode. The relay UE may use multiple access schemes such as 5G RAT, LTE, WiFi, and fixed broadband. Service continuity plays a key role when changing from one relay UE to another or to the direct network connection mode. In addition, the 5G system is expected to support the battery consumption optimization of relay UEs.

2.1. IoT and Wearables

IoT devices with a very long expected battery lifetime and wearables with other cellular-connected devices in their proximity would especially benefit from short D2D links. Motivated by this, 3GPP opened a release 15 study item "Study on Further Enhancements to LTE Device to Device, UE to Network Relays for IoT and Wearables" [42]. The primary objective of the study was to improve the power efficiency of the remote UEs (IoT devices and wearables) by allowing them to form a D2D connection with a UE who is willing to act as a relay [40]. Enhancements were planned to release 13 UE-to-network relaying to support end-to-end security and QoS as well as efficient path switching between conventional and D2D air interfaces. In addition, the needed changes for sidelink were studied to provide a reliable D2D communication link for low cost and low power IoT devices.

The study considered a diverse group of scenarios that could benefit from UE-to-network relaying. From the coverage point of view, the remote UE could be located within the cell, out of cell, or can be operating in the coverage-enhanced mode [40]. As cellular IoT devices mainly reach enhanced coverage by a high number (up to 2048) of repeated transmissions [43], the power efficiency gain of using short D2D links with minimal repetitions is obvious in this scenario. Relaying using the sidelink can be bi- or uni-directional, as shown in Figure 2. Bidirectional relaying is more straightforward to implement with minimal signaling from the eNB. However, bidirectional relaying over sidelink requires UL waveform reception capabilities for the remote UEs. This would mean implementing a UL receiver for low-cost IoT devices, which may not be feasible from the device cost point of view. Thus, many of the open issues in D2D relaying for IoT are related to the question, how to efficiently implement mandatory functionalities, such as discovery, for unidirectional relaying.

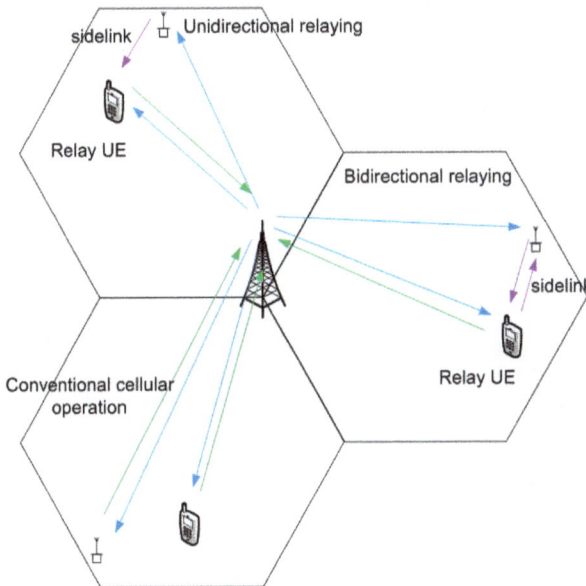

Figure 2. Device-to-device (D2D) relaying variants for cellular Internet of Things (IoT) devices; UE: user equipment.

As a result of the 3GPP study, a relaying architecture was proposed. Relaying is done above the radio link control (RLC) layer, i.e., the RLC and lower layers are terminated at the D2D link and higher layers at the remote UE and the eNB [33]. Several solutions for paging and system information transfer for remote UEs as well as path switch and group handover enhancements were also proposed. These layer 2 studies mostly assumed the feasibility of bidirectional relaying; the impact of unidirectional relaying was not fully analyzed in the study item. For example, the discovery procedure for the unidirectional relaying case with remote UEs only capable of receiving downlink (DL) signals was still left open. Another aspect in the 3GPP study was to study the required enhancements to sidelink physical layer operation. The target was to also enable the sidelink support for low-cost UEs with a limited bandwidth of one (narrowband IoT) or six (LTE-M) physical resource blocks (PRBs) and potentially with no sidelink reception capabilities [33]. Enhancements were proposed to the synchronization procedure such that the relay UE can act as a synchronization source for the remote UEs. Also, the needed enhancements for the support of unicast communications over the sidelink were identified and proposed for resource allocation, semi-persistent scheduling, power control, measurements and feedback for link adaptation. Based on the performance evaluation results presented in [33], especially the adaptive modulation and coding together with the adaptive number of sidelink transmissions provided a significant energy efficiency gain for the remote UEs.

There are still several open issues regarding D2D and UE relaying for cellular IoT. From the research point of view, the effect on the cell energy efficiency and the battery life-time for all involved devices has not been thoroughly studied. It is clear that with UE relaying, the devices willing to operate as relays consume more power than the remote UEs. However, the device power consumption model used in [33] was rather simplified and no clear view on the spatial distribution of the power consumption was achieved. The 3GPP has plans to continue the normative work on bringing the relaying support for cellular IoT and wearables into standards. Currently, the corresponding work item has been proposed, but it is yet unclear whether the work will take place in release 15 or 16 [44]. D2D communication support in different 3GPP releases is depicted in Figure 3.

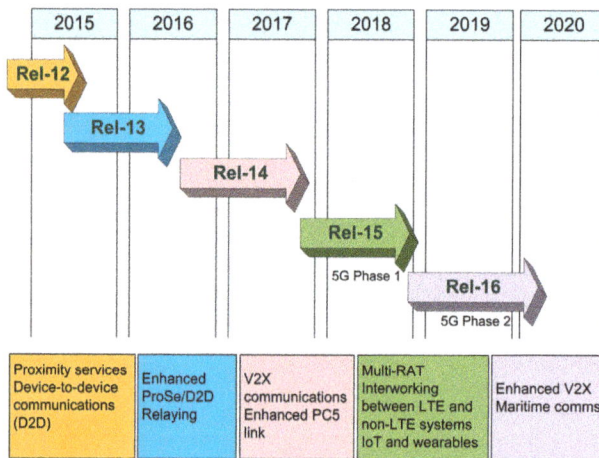

Figure 3. D2D communications support in 3rd Generation Partnership Project (3GPP) releases; V2X: Vehicle-to-Everything; Rel: Release; RAT: Radio Access Technology.

2.2. Vehicle-to-Everything (V2X) and Maritime Communications

Another important area for D2D communications is vehicular communications or V2X communications that can be divided into three areas, namely vehicle-to-vehicle (V2V), vehicle-to-infrastructure (V2I), and vehicle-to-network (V2N) [9]. The V2V and V2I communications towards the other vehicles and roadside units (RSU) are handled through the PC5 interface in 3GPP networks. Connectivity to the network and the cloud (V2N) goes through the Uu interface (Air interface). V2X communications is included first time in Release 14.

Enhanced support for V2X services (eV2X) in 3GPP release 15 will include safety-related V2X scenarios, such as automated and remote driving and platooning, where vehicles form a platoon or a line travelling together [45]. It will also enable extended sensors where vehicles could exchange sensor information locally. A relevant aspect of advanced V2X applications is the level of automation (LoA), which reflects the functional aspects of the technology and affects the system performance requirements. The levels of automation are defined as: 0—No Automation, 1—Driver Assistance, 2—Partial Automation, 3—Conditional Automation, 4—High Automation, 5—Full Automation.

At lower automation levels a human operator is primarily responsible for monitoring the driving environment, whereas in higher layers an automated system is responsible for operations. Similar types of work are going on in the development of automated drones and autonomous and remote-controlled ships [8]. Currently 3GPP is considering and developing systems specifically for maritime communications for release 16 and beyond to support the needs of future maritime users [46]. One of the requirements of this "LTE-Maritime" system is to support 100 km coverage. It will also support the interworking between the 3GPP system and the existing/future maritime radio communication system for the seamless service of voice communication and data communication between users ashore and at sea or between vessels at sea.

3. System Model and D2D Use Cases for Combined LTE/5G and WiFi

Figure 4 presents our high-level system model for D2D communications in a 5G network. There are many types of users that are connected to the base stations using cellular interfaces. Nodes can also communicate directly using D2D communication links between nodes that are in proximity to each other. Direct links between user devices such as phones and laptops may use several RATs, including 3GPP evolution, as described in Section 2, Bluetooth, or WiFi standards. Cars also use a dedicated 802.11p standard in the intelligent transport system (ITS) band in 5.9 GHz for V2X communications. In the future, autonomous and remote-controlled ships will also use more and more ship-to-ship communications, possibly also radios specifically developed for these purposes. Both in the V2X communications among cars and in maritime communications, integrated 5G satellite-terrestrial systems will be needed [8,9].

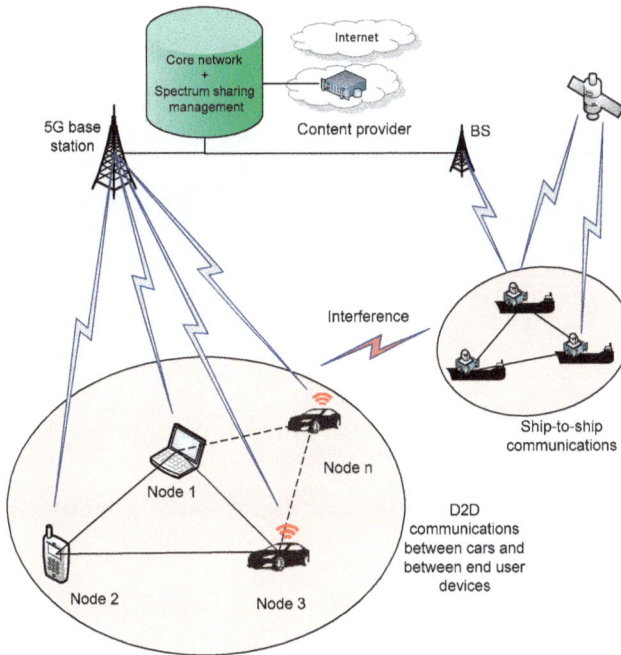

Figure 4. High-level system model for D2D communications in 5G; BS: Base station.

The system has a connection to the Internet and the connectivity provider to make all the required services available to the end users. The 5G core supports seamless cooperation between different RATs and the terrestrial and satellite segments. It also enables QoS management of data transmission e.g., by dedicating part of the resources to applications with higher priority. There could even be end-to-end network slices dedicated to autonomous driving and other use cases so that QoS requirements can be met in any circumstances via proper resource allocation and isolation mechanisms. Network virtualization and slicing techniques enable different operators to share network resources with other (virtual) operators and to provide end-to-end connectivity across operator boundaries.

In addition to network management with the core network, the 5G networks will also use spectrum sharing technologies to utilize available radio resources as efficiently as possible. We assume a licensed spectrum access (LSA) approach, where the incumbent operators are required to provide a priori information about their spectrum use over the area of interest to the database. They tell explicitly where, when, and which parts of the frequency bands are available for the secondary use. This most probably requires a third party to operate the LSA system, since operators are often not willing to share the information about their spectrum use with other spectrum users.

Let us now look at the simplified model for the analysis that is presented in Figure 5. The model is based on the high-level system model described above. Wireless mobile users are connected to the base station using the LTE interface. There are N nodes in the network. We assume that links L_{12} (between Node 1 and Node 2), L_{13}, L_{23}, L_{3n} can be either LTE or WiFi links. Only user equipment such as phones, tablets, and laptops are used as nodes in the network. Link attenuations between the base station and the user equipment are assumed to be equal, as well as the direct links between nodes. All the links between the user equipment and the base station are using 3GPP interfaces.

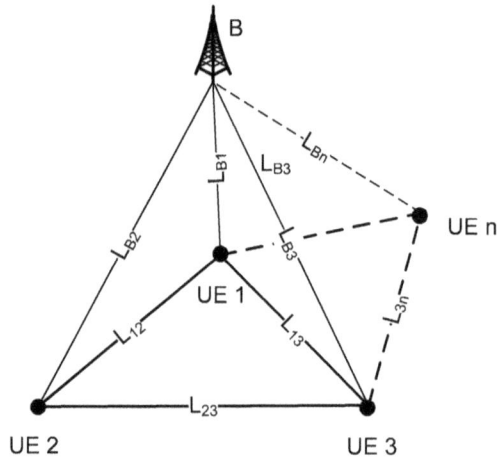

Figure 5. Simplified model for analysis. LB1: Link between the base station and the UE 1; LB2: Link between the base station and the UE 2; LB3: Link between the base station and the UE 3; LBn: Link between the base station and the UE n; L12: D2D link between UE 1 and UE 2; L13: D2D link between UE 1 and UE 3; L23: D2D link between UE 2 and UE 3; L3n: D2D link between UE 3 and UE n.

D2D communication is controlled by the base station, which enables interference management and assures QoS to the end users. Nodes can form a cluster around the cluster head which may be the only node discussing with the base station. In order to estimate the power consumption in the depicted system model both from the user device and the base station perspectives, we need to define practical use cases for analysis. Based on Figure 5, we can define several different use cases for delivering Internet data or some other data from the content provider that certain node(s) want to access through the base station. Five different cases are described in the following as [30,31]:

(1) **Case 1:** The base station sends the data directly to the requesting node(s).
(2) **Case 2:** Nodes with social ties form a cluster. The base station sends the data to the cluster head that relays the data to other users over WiFi. The data (such as recently popular YouTube videos) is cached in the cluster head for some time in order to serve requesting nodes directly.

(3) **Case 3:** The base station sends the data to the cluster head that relays the data to requesting nodes over LTE.

(4) **Case 4:** The base station sends $1/N$ of the required packets to N different nodes requesting the same data (e.g., certain content in Facebook shared among friends). Different parts are sent to different users and the missing parts are shared using D2D connections among nodes over WiFi.

(5) **Case 5:** Same as case 4, but the sharing is done using an LTE interface.

4. Power Consumption Models

4.1. LTE Base Station Model

The majority of the energy in wireless networks is consumed in the base stations, also in the defined cooperative scenarios. From the base station point of view, it is crucial to study the supply power consumption rather than radio frequency (RF) transmission power to see the total effect. Supply power consumption P_{sup} for a single RF chain showing the relation between supply power and RF transmission power P_{tx} is [21]:

$$P_{sup} = \begin{cases} P_0 + \Delta_p P_{tx}, & 0 < P_{tx} < P_{max} \\ P_{sleep}, & P_{tx} = 0 \end{cases} \tag{1}$$

where P_0 is the minimum active power consumption, Δ_p is a linear transmission dependence factor, and P_{sleep} is the power consumption in the sleep mode. When there are N_{trx} RF chains included, the total supply power consumption P_{tot} is

$$P_{tot} = N_{trx} \times P_{sup} \tag{2}$$

Measured parameter values of LTE base stations (macro, remote radio head, micro, pico, femto) can be found from [22]. The values are summarized in Table 1. The model and the values are based on commercially-available base stations, providing sufficient foundation for our energy estimations. We adopt this model since it is simple, based on vigorous measurements, and easy to use in the analysis. We note that there are also other models recently published, such as in [47], where a general conclusion is drawn as: "Modeling a linear dependence between the emitted power and the energy consumption, as well as between the traffic volume and the energy consumption, is a very good approximation, and it is strongly confirmed by real data".

Table 1. Base station power consumption parameters. Data from [22]; BS: Base station; N_{trx}: number of radio frequency (RF) chains; P_{max}: Maximum transmission power; P_0: Minimum active power consumption; P_{sleep}: Power consumption in the sleep mode.

BS Type	N_{trx}	P_{max} (W)	P_0 (W)	Δ_p	P_{sleep} (W)
Macro	6	39.8	130.0	4.7	75.0
Remote radio head	6	20	84.0	2.8	56.0
Micro	6	6.3	56.0	2.6	39.0
Pico	2	0.13	6.8	4.0	4.3
Femto	2	0.05	4.8	8.0	2.9

4.2. Model for LTE User Device

The power consumption (mW) when receiving data in a connected state is estimated as [18]:

$$P_{rx} = P_{on} + P_{rxBB}(R_{rx}) + P_{rxRF}(S_{rx}) + \beta_{rx} \tag{3}$$

where P_{on} is the power consumption when the cellular subsystem is active, β_{rx} is the additional power consumption of a receiver being active. Parameter P_{rxRF} defines radio frequency (RF) block power consumption that is dependent on the received power S_{rx} and P_{rxBB} is the baseband power consumption, dependent on the received data rate R_{rx}. These parameters are given as

$$P_{rxRF} = \begin{cases} -0.04 \times S_{rx} + 24.8, & S_{rx} \leq -52.5 \text{ dBm} \\ -0.11 \times S_{rx} + 7.86, & S_{rx} > -52.5 \text{ dBm} \end{cases}$$

$$P_{rxBB} = 0.97 R_{rx} + 8.16$$

Equivalent power consumption (mW) when transmitting data in the connected state is given as:

$$P_{tx} = P_{on} + P_{txBB}(R_{tx}) + P_{txRF}(S_{tx}) + \beta_{tx} \tag{4}$$

where same parameters are defined for the transmitter side, respectively. Transmission power S_{tx} primarily affects the RF block power consumption:

$$P_{txRF} = \begin{cases} 0.78 \times S_{tx} + 23.6, & S_{tx} \leq 0.2 \text{ dBm} \\ 17.0 \times S_{tx} + 45.4, & 0.2 \text{ dBm} < S_{tx} \leq -11.4 \text{ dBm} \\ 5.90 \times S_{tx}^2 - 118 \times S_{tx} + 1195, & 11.4 \text{ dBm} < S_{tx} \end{cases}$$

The data rate does not affect baseband power consumption in the uplink, i.e., P_{txBB} is constantly 0.62 mW. Other parameters are P_{on} = 853 mW, β_{rx} = 25.1 mW and β_{tx} = 29.9 mW.

4.3. WiFi Power Consumption Models

The power consumption model for LTE and WiFi 802.11g air interfaces has linear dependency on the data rate in measurements done in [19], as shown in the following. Power consumption (mW) when receiving data is estimated as

$$P_{rx} = \alpha_{rx} R_{rx} + \beta \tag{5}$$

The power consumption (mW) when transmitting data is estimated as

$$P_{tx} = \alpha_{tx} R_{tx} + \beta \tag{6}$$

The parameters α_{rx} and α_{tx} are linear scaling factors for reception and transmission, R_{rx} is the received data rate, R_{tx} is the transmitted data rate and β is the basic power consumption in the active mode. Based on several references, parameters for these models are given in Table 2. It can be seen that the older air interfaces behave according to Equations (5) and (6), including the LTE device model in [19] and the 802.11g model in the same paper. The more recent 802.11n model that was defined in [30] based on measurements reported in [20] is quite flat.

Table 2. Power consumption parameters of different long term evolution (LTE) and WiFi models; α_{rx}: linear scaling factor for reception; α_{tx}: linear scaling factor for transmission; β: basic power consumption in the active mode.

Ref.	Air Interface	ff_{rx} (mW/Mbps)	ff_{tx} (mW/Mbps)	β (mW)
[19]	LTE	51.97	438.39	1288.04
	WiFi, 802.11g	137.01	283.17	132.86
[20,30]	WiFi, 802.11n	6	4	β_{rx} = 450, β_{tx} = 980
[32]	802.11ac	~2100 mW *	~2500mW *	287
[32]	802.11ad	~2100 mW *	~2000 mW *	1938

* Over a large bit rate range the power consumption is quite flat in recent 802.11ac and ad interfaces.

The most recent 802.11ac and 802.11ad measurements given in [32] show that both receiver power consumption and transmitter power consumption are almost flat, regardless of the bit rate. The basic power consumption is much lower in 802.11ac, but the 802.11ad interface always consumes a lot of energy when it is active. There is no big difference when receiving or transmitting data compared to the basic power consumption according to [32]. However, the results indicate that with the latest models, the best option for energy saving is to send the data with the best possible data rate in order to be able spend more time in the basic power consumption mode.

5. Performance Analysis

5.1. Power Consumption of the End User Device

Mathematically, the power consumption within the cooperative network in defined use cases can be given as follows: In case 1 the end user devices are only receiving the data using the LTE interface. Thus, the power consumption in this reference case is

$$P_{tot} = N \times P_{rx, \text{ LTE}}(R)$$ (7)

where $P_{rx, \text{ LTE}}$ is the received signal power for a signal coming from the base station. In case 2, one node is receiving the data over the LTE link and transmits the data over WiFi to N-1 users, i.e.,

$$P_{tot} = P_{rx, \text{ LTE}}(R) + P_{tx, \text{ WiFi}}(R) + (N-1) \times P_{rx, \text{ WiFi}}(R)$$ (8)

In case 3, the same transmissions are conducted over the LTE interface. Thus, the total power consumption is

$$P_{tot} = P_{rx, \text{ LTE}}(R) + P_{tx, \text{ LTE\_D2D}}(R) + (N-1) \times P_{rx, \text{ LTE\_D2D}}(R)$$ (9)

where $P_{tx, \text{ LTE\_D2D}}$ is the transmission power consumption of a UE and $P_{rx, \text{ LTE\_D2D}}$ is the received power consumption for a D2D signal. R is the required data rate over the link. In cases 4 and 5, the data rate is divided into multiple R/N rate streams that are then combined at the requesting node(s). In case 4, the total power consumption is

$$P_{tot} = N \times P_{rx, \text{ LTE}}(R/N) + N \times P_{tx, \text{ WiFi}}(R/N) + N \times P_{rx, \text{ WiFi}}(R - R/N)$$ (10)

and in case 5 it is

$$P_{tot} = N \times P_{rx, \text{ LTE}}(R/N) + N \times P_{tx, \text{ LTE\_D2D}}(R/N) + N \times P_{rx, \text{ LTE\_D2D}}(R - R/N)$$ (11)

The power consumption of the cluster head is given in Equations (8) and (9) by excluding the last term in the equation. In cases 4 and 5, the power consumption is equally shared between the nodes.

5.2. Energy Consumption of a Base Station

Resource allocations in the time and frequency domains in the defined use cases are presented in Figure 6. Cooperation leads to a shorter active transmission period of the base station in all co-operative scenarios. The figure shows an example with two nodes (UEs) but the same model can be easily generalized to N users. The energy required for the transmission of data is the integral of the power consumption $P(t)$ of the air interface over time

$$E = \int_{t_0}^{t_0+T} P(t)dt$$ (12)

where the transmission duration T is dependent on the transmission size D and data rate R of the used air interface. We can now define the energy consumption for all defined cases as follows.

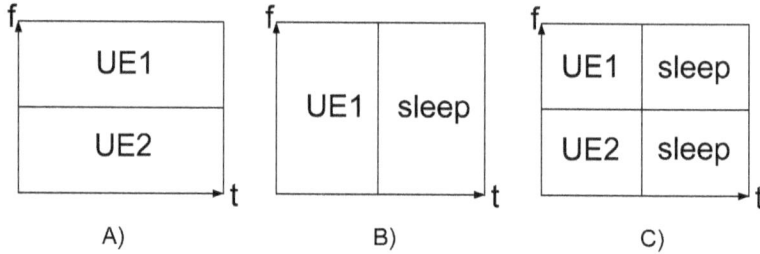

Figure 6. Resource allocation from the BS perspective assuming two mobile devices. (**A**) Case 1, (**B**) Cases 2 and 3, and (**C**) Cases 4 and 5.

Case 1: Normal cellular case, data sent independently to N users. According to Equation (2) energy consumption is

$$E = N_{\text{trx}} \times (P_0 + \Delta_p P_{\text{tx}}) \times (D/R) \tag{13}$$

Case 2 and Case 3 look the same from the base station perspective, since it sends all the data to a single relay. Clear energy savings are achieved especially if the same data is of interest to multiple users in a D2D enabled network. Energy consumption is now defined as

$$E = N_{\text{trx}} \times (P_0 + \Delta_p P_{\text{tx}}) \times \frac{1}{N}\left(\frac{D}{R}\right) + P_{\text{sleep}} \frac{N-1}{N}\left(\frac{D}{R}\right) \tag{14}$$

which means that the base station is able to reduce its active transmission time to one Nth of the time when compared with the Case 1 and then spend rest of the time in the sleep mode.

Again, Case 4 and Case 5 are the same from the base station perspective. Since the data is divided into independent pieces, the total amount of data transmitted by the base station is actually the same as in Case 2 and Case 3. Assuming that separating the interesting data to independent pieces does not consume significant amount of energy, we can use the same model for the base station power consumption as in Equation (14).

6. Results

6.1. Power Consumption of End User Devices

Figures 7–9 show power consumption results with the defined power consumption models from the end user perspective. All used WiFi and LTE models are applicable to out-of-band D2D communication scenarios. In addition, the LTE models are applicable to in-band overlay D2D where D2D links use dedicated resources. The power consumption of the total D2D network, as well as the power consumption of the cluster head of a network in each case, is given in Figure 7 for a cluster size of $N = 4$ nodes using the Huang model [19] for the LTE and WiFi interfaces. It is seen that with the low throughput values it is best that only the cluster head actively receives the data from the LTE base station. Then it uses WiFi for relaying the data to requesting users. However, it can be seen that from the cluster head perspective this is the second most power consuming option and thus there might be a need to change the cluster head from time to time in order to prevent it draining the battery completely. When the higher throughput >6 Mbps is required, the most power efficient option from the end user perspective is to receive all the data directly from the base station.

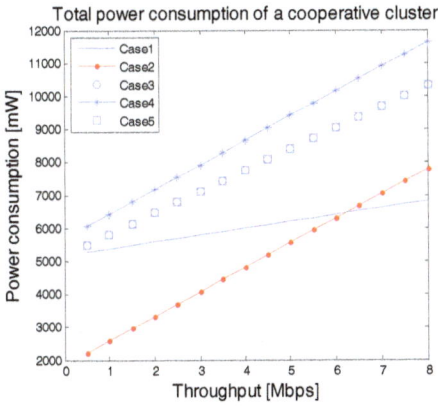

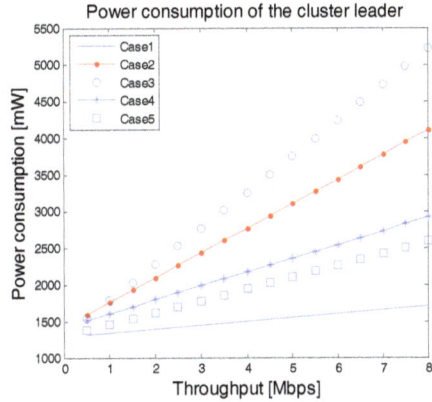

Figure 7. Power consumption with the Huang LTE and WiFi models, 4 nodes; whole network (**left**) and the cluster leader (**right**).

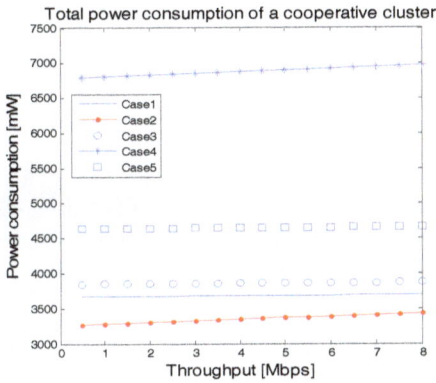

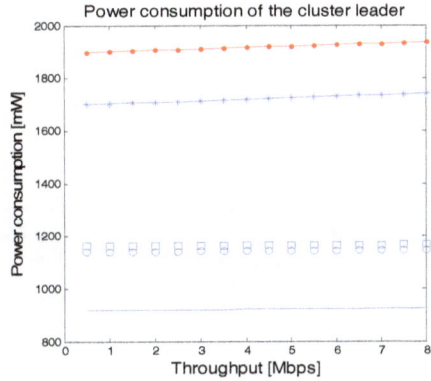

Figure 8. Results with the Lauridsen LTE and the 802.11n WiFi; Power consumption of a whole network (**left**) and the cluster leader (**right**).

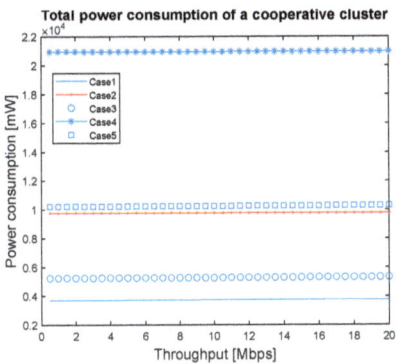

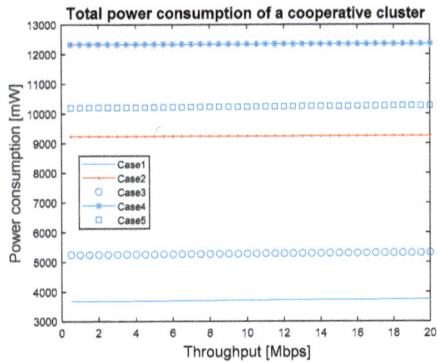

Figure 9. Results with the Lauridsen LTE and the 802.11ac (**left**) and 802.11ad (**right**) WiFi, whole network considered.

When the Lauridsen model is adopted for LTE and 802.11n for WiFi, the observations are a bit different, as seen in Figure 8. We have assumed $S_{rx} = -50$ dBm and $S_{tx} = 10$ dBm for a D2D LTE link. The total power consumption in case 2 with a higher number of nodes is even more advantageous due to the lower power consumption of the WiFi. Case 4 demands the active operation of both LTE and WiFi interfaces. This is not good from the power consumption point of view due to the static part of the power consumption that comes from keeping the air interface active, i.e., β in Equations (5) and (6). Thus, the latest power consumption models propose that dividing the data into smaller streams and changing the missing packets over the air is not efficient due to the simultaneous use of several active interfaces. WiFi relaying is a good option up to 20 Mbps data rate. However, also in this case, one has to take care that the cluster head is changed from time to time in a mobile network to keep all the nodes alive for longer periods of time.

The situation is quite similar when the 802.11ac and 802.11ad WiFi models are adopted as seen in Figure 9. The results cover the whole network and show that with the latest radios, where the power consumption is static regardless of the data rate, the best option is to use LTE alone. Either the conventional cellular operation or relaying with LTE are the best choices. This is due to the high power consumption of WiFi models with any data rate. An active WiFi interface consumes a lot of power. WiFi could be used to enhance the data rate of the devices if very high data rate services were needed.

The used models are applicable both to LTE and LTE-advanced systems. Only some parameter updates are needed e.g., regarding the power model given in the Equation (1). e.g., release 12 equipment in our lab uses the old HW and only the SW is updated in the base station compared to the older releases. Power consumption is affected by the software as well, but the LTE-A base station power consumption can be described with the same model due to the slow evolution of the devices.

6.2. Base Station Energy Consumption in D2D Networks

The energy consumption of cooperative scenarios from the base station perspective is the same for all depicted D2D scenarios. Thus, we compare here conventional cellular operation with the cooperative scenario as a function of number of nodes in a D2D network. We adopt the energy consumption metric J/bit [22] that focuses on the amount of energy spent per delivered bit and is hence an indicator of network bit delivery efficiency.

We assume an average bit rate of 10 Mbit/s in the following figures and use the energy consumption models of Equations (13) and (14). The transmission power P_{tx} is set according to the P_{max} values in Table 2. The results presented in Figure 10 for a macro base station show that with this data rate conventional cellular transmission consumes roughly 0.3 J/kbit, whereas the cooperation clearly reduces the energy consumption by sharing the load among cooperative nodes. The effect is the largest with a few additional cooperative nodes, three nodes already lead to 50% energy savings. When the number of nodes is increased to more than 10 nodes, the energy consumption of a base station is around 0.1 J/kbit which means that the base station is able to serve the requesting nodes with one third of the original energy. This is a significant improvement in the energy efficiency.

When the cell size is smaller, the energy efficiency improvement is smaller, as can be seen in Figures 11 and 12. Still, even with the small cell base stations the energy reduction is around 40%, which is significant saving already with a few requesting nodes. The results suggest that cooperative D2D data dissemination approaches are good for cellular network energy efficiency. The gain is dependent on the D2D link quality, and with poor D2D links the energy savings would be smaller.

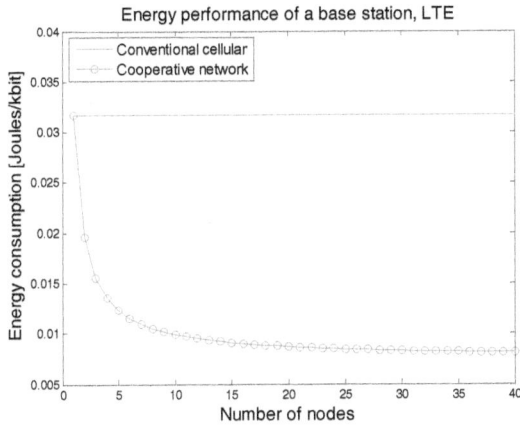

Figure 10. Energy consumption of a macro base station.

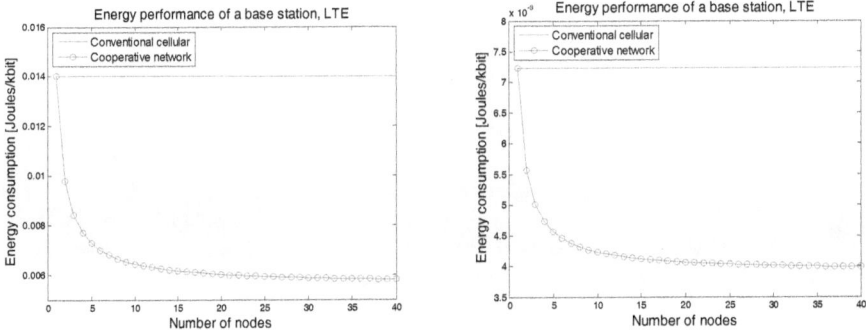

Figure 11. Energy consumption of a remote radio head (**left**) and a micro base station (**right**).

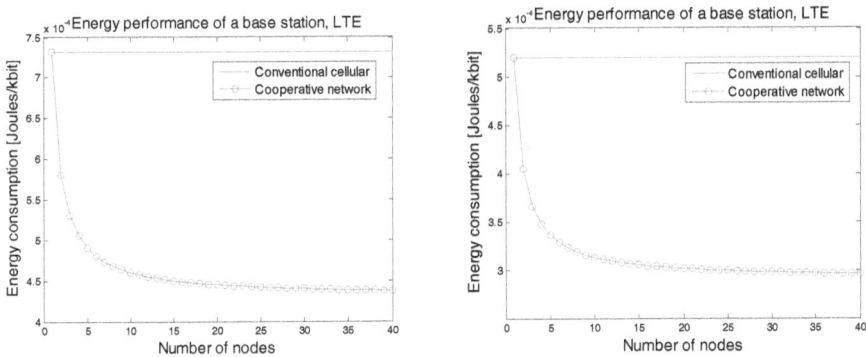

Figure 12. Energy consumption of pico (**left**) and femto (**right**) base stations.

7. Recommendations

Based on the conducted analysis, the following recommendations can be made for network deployment and operations:

(1)   When the power consumption is dependent on the data rate (as in Figure 7), the aim should be to find the sweet spots or data rate regions to use different air interfaces. In multi-RAT 5G networks this would mean analysis of all other radio interface options than the ones analyzed in this paper. However, the most important ones currently are the LTE and WiFi.

(2)   With the latest WiFi and LTE models, the best option for cooperative data delivery is to select a relay and then use LTE for D2D transmissions. WiFi is a good option only for very high data rates.

(3)   The base station results show that D2D transmission brings the largest gains in macro cells, up to 70% energy reductions. Small cells are already more energy efficient, but energy savings can still be significant, even 40% in the case of femto cells. Thus, it is recommended to use direct communication between devices in all cellular networks regardless of the type of base station.

(4)   The best option for energy saving in D2D communications using the latest LTE and WiFi models is to send the data with the best possible data rate.

(5)   There is clearly room for the creation of novel models with the latest 3GPP releases after rigorous power consumption measurement work. The measurements should specifically consider D2D measurements with the release 12 and beyond devices.

8. Conclusions

Energy efficiency is an important factor in 5G and beyond networks and one of the drivers in the adoption of D2D technology. This paper has reviewed the potential application areas including IoT, wearables, and automated driving and reviewed the current status of D2D technology in 3GPP standardization. In addition, we have analyzed D2D-enhanced cellular networks both from the base station and from the end user perspectives. The analysis is conducted with several different measurement-based LTE and WiFi models. The results show that significant energy reductions can be achieved with all types of base stations, including macro, pico, and femto base stations. The results also suggest that in order to minimize power consumption, the devices should minimize the number of active radio interfaces and use the best possible data rates. In our system model this means that either a LTE or WiFi interface is active in a single device at a given time instant. WiFi could be used to support very high data rate services. If there is no need for that, one should keep only the LTE interface active in order to save power. An interesting future topic could be to study the effect of mobility in energy consumption. This would create new challenges e.g., due to frequent handovers in a multi-RAT network. In addition, adaptive power control could be included in the analysis to gain a more detailed understanding e.g., on the effect of UL transmissions.

Acknowledgments: This work was supported by the CORNET project, partly funded by Tekes, the Finnish Funding Agency for Innovation.

Author Contributions: Marko Höyhtyä performed the experiments and analyzed the data. He was also the main author of the paper. Olli Apilo wrote a major part of the 3GPP section, especially regarding the IoT and wearables. Mika Lasanen commented on and supported work throughout the paper.

Conflicts of Interest: The authors declare no conflict of interest.

References

1.   Adachi, T.; Nakagawa, M. A study on channel usage in a cellular Ad-Hoc united communication system. *IEICE Trans. Commun.* **1998**, *81*, 1500–1507.

2.   3rd Generation Partnership Project (3GPP). TR 36.843 V12.0.1, Study on LTE Device to Device Proximity Services, Release 12. March 2014. Available online: http://www.3gpp.org/ftp/Specs/archive/36_series/36.843/36843-c01.zip (accessed on 27 December 2017).

3.   Asadi, A.; Wang, Q.; Mancuso, V. A survey on device-to-device communication in cellular networks. *IEEE Commun. Surv. Tutor.* **2014**, *16*, 1801–1819. [CrossRef]

4.   Usman, M.; Gebremariam, A.A.; Raza, U.; Granelli, F. A software-defined device-to-device communication architecture for public safety applications in 5G networks. *IEEE Access* **2015**, *3*, 1649–1654. [CrossRef]

5. Gallo, L.; Härri, J. Unsupervised long-term evolution device-to-device: A case study for safety critical V2X communications. *IEEE Veh. Technol. Mag.* **2017**, *12*, 69–77. [CrossRef]

6. Bello, O.; Zeadally, S. Intelligent device-to-device communication in the Internet-of-Things. *IEEE Syst. J.* **2016**, *10*, 1172–1182. [CrossRef]

7. Alam, M.M.; Arbia, D.B.; Hamida, E.B. Research trends in multi-standard device-to-device communication in wearable wireless networks. In Proceedings of the International Conference on Cognitive Radio Oriented Wireless Networks, Doha, Qatar, 21–23 April 2015.

8. Höyhtyä, M.; Huusko, J.; Kiviranta, M.; Solberg, K.; Rokka, J. Connectivity for autonomous ships: Architecture, use cases, and research challenges. In Proceedings of the 8th International Conference on ICT Convergence, Jeju Island, Korea, 18–20 October 2017.

9. Höyhtyä, M.; Ojanperä, T.; Mäkelä, J.; Ruponen, S.; Järvensivu, P. Integrated satellite-terrestrial systems: Use cases for road safety and autonomous ships. In Proceedings of the 23rd Ka and Broadband Communications Conference, Trieste, Italy, 16–19 October 2017.

10. Cisco Systems, Inc. *Cisco Visual Networking Index: Global Mobile Data Traffic Forecast Update, 2016–2021 White Paper*; Cisco: San Jose, CA, USA, 7 February 2017; Available online: https://www.cisco.com/c/en/us/solutions/collateral/service-provider/visual-networking-index-vni/mobile-white-paper-c11-520862.pdf (accessed on 22 December 2017).

11. Chen, T.; Yang, Y.; Zhang, H.; Kim, H.; Horneman, K. Network energy saving technologies for green wireless access networks. *IEEE Wirel. Commun.* **2011**, *18*, 30–38. [CrossRef]

12. Zhang, J.; Wang, Z.-J.; Quan, Z.; Yin, J.; Chen, Y.; Guo, M. Optimizing power consumption of mobile devices for video streaming over 4G LTE networks. *Peer-to-Peer Netw. Appl.* **2017**. [CrossRef]

13. Militano, L.; Iera, A.; Molinaro, A.; Scarcello, F. Energy-saving analysis in Cellular-WLAN cooperative scenarios. *IEEE Trans. Veh. Technol.* **2014**, *63*, 478–484. [CrossRef]

14. Rao, J.B.; Fapojuwo, A.O. A survey of energy efficient resource management techniques for multicell cellular networks. *IEEE Commun. Surv. Tutor.* **2014**, *16*, 154–180. [CrossRef]

15. Mämmelä, A.; Anttonen, A. Why will computing power need particular attention in future wireless devices? *IEEE Circuits Syst. Mag.* **2017**, *17*, 12–26. [CrossRef]

16. Caire, G.; Taricco, G.; Biglieri, E. Optimum power control over fading channels. *IEEE Trans. Inf. Theory* **1999**, *45*, 1468–1489. [CrossRef]

17. Höyhtyä, M.; Mämmelä, A. A unified framework for adaptive inverse power control. *EURASIP J. Wirel. Commun. Netw.* **2016**, *2016*, 41. [CrossRef]

18. Lauridsen, M.; Noël, L.; Sørensen, T.B.; Mogensen, P. An empirical LTE smartphone power model with a view to energy efficiency evolution. *Intel Technol. J.* **2014**, *18*, 172–193.

19. Huang, J.; Qian, F.; Gerber, A.; Mao, Z.M.; Sen, S.; Spatscheck, O. A close examination of performance and power characteristics of 4G LTE networks. In Proceedings of the 10th Mobile Systems, Applications, and Services, Low Wood Bay, UK, 25–29 June 2012.

20. Saha, S.K.; Deshpande, P.; Inamdar, P.P.; Sheshadri, R.K.; Koutsonikolas, D. Power-throughput tradeoffs of 802.11n/ac smartphones. In Proceedings of the 2015 IEEE Computer Communications (INFOCOM), Hong Kong, China, 26 April–1 May 2015.

21. Holtkamp, H.; Auer, G.; Bazzi, S.; Haas, H. Minimizing base station power consumption. *IEEE J. Sel. Areas Commun.* **2014**, *32*, 297–306. [CrossRef]

22. EARTH Project Deliverable D2.3. Energy Efficiency Analysis of the Reference Systems, Areas of Improvements and Target Breakdown. January 2012. Available online: http://cordis.europa.eu/docs/projects/cnect/3/247733/080/deliverables/001-EARTHWP2D23v2.pdf (accessed on 22 December 2017).

23. Zuo, X.; Iamnitchi, A. A survey of socially aware peer-to-peer systems. *ACM Comput. Surv.* **2016**, *49*, 9. [CrossRef]

24. Gupta, M.; Jha, S.C.; Koc, A.T.; Vannithamby, R. Energy impact of emerging mobile Internet applications on LTE networks: Issues and solutions. *IEEE Commun. Mag.* **2013**, *51*, 90–97. [CrossRef]

25. Koc, A.T.; Jha, S.C.; Vannithamby, R.; Torlak, M. Device power saving and latency optimization in LTE-A networks through DRX configuration. *IEEE Trans. Wirel. Commun.* **2014**, *13*, 2614–2625.

26. Larsson, E.G.; Edfors, O.; Tufvesson, F.; Marzetta, T.L. Massive MIMO for next generation wireless systems. *IEEE Commun. Mag.* **2014**, *52*, 186–195. [CrossRef]

27. Lei, L.; Zhong, Z.; Lin, C.; Shen, X. Operator controlled device-to-device communications in LTE-Advanced networks. *IEEE Wirel. Commun.* **2012**, *19*, 96–104. [CrossRef]
28. Chih-Lin, I.; Rowell, C.; Han, S.; Xu, Z.; Li, G.; Pan, Z. Toward green and soft: A 5G perspective. *IEEE Commun. Mag.* **2014**, *52*, 66–73.
29. Zhang, H.; Huang, S.; Jiang, C.; Long, K.; Leung, V.C.M.; Poor, H.V. Energy efficient user association and power allocation in millimeter-wave-based ultra dense networks with energy harvesting base stations. *IEEE J. Sel. Areas Commun.* **2017**, *35*, 1936–1947. [CrossRef]
30. Höyhtyä, M.; Mämmelä, A.; Celentano, U.; Röning, J. Power-efficiency in social-aware D2D communications. In Proceedings of the European Wireless 2016 22th European Wireless Conferences, Oulu, Finland, 18–20 May 2016.
31. Höyhtyä, M.; Mämmelä, A. Energy-efficiency in social-aware D2D networks: A base station perspective. In Proceedings of the 2016 Advances in Wireless and Optical Communications (RTUWO), Riga, Latvia, 3–4 November 2016.
32. Saha, S.K.; Siddiqui, T.; Koutsonikolas, D.; Loch, A.; Widmer, J.; Sridhar, R. A detailed look into power consumption of commodity 60 GHz devices. In Proceedings of the 2017 IEEE 18th International Symposium on A World of Wireless, Mobile and Multimedia Networks (WoWMoM), Macau, China, 12–15 June 2017.
33. 3rd Generation Partnership Project (3GPP). TR36.746 Study on Further Enhancements to LTE Device to Device (D2D), UE to Network Relays for IoT (Internet of Things) and Wearables, V2.0.1. October 2017. Available online: http://www.3gpp.org/ftp//Specs/archive/36_series/36.746/36746-201.zip (accessed on 27 December 2017).
34. 3rd Generation Partnership Project (3GPP). TS29.214 Policy and Charging Control over Rx Reference Point, V14.3.0. March 2017. Available online: http://www.3gpp.org/ftp//Specs/archive/29_series/29.214/29214-e30.zip (accessed on 27 December 2017).
35. Rost, P.; Mannweiler, C.; Michalopoulos, D.S.; Sartori, C.; Sciancalepore, V.; Sastry, N.; Holland, O.; Tayade, S.; Han, B.; Bega, D.; et al. Network slicing to enable scalability and flexibility in 5G mobile networks. *IEEE Commun. Mag.* **2017**, *55*, 72–79. [CrossRef]
36. Zhang, H.; Liu, N.; Chu, X.; Long, K.; Aghvami, A.-H.; Leung, V.C.M. Network slicing based 5G and future mobile networks: Mobility, resource management, and challenges. *IEEE Commun. Mag.* **2017**, *55*, 138–145. [CrossRef]
37. Lin, X.; Andrews, J.G.; Ghosh, A.; Ratasuk, R. An overview of 3GPP device-to-device proximity services. *IEEE Commun. Mag.* **2014**, *52*, 40–48. [CrossRef]
38. Mach, P.; Becvar, Z.; Vanek, T. In-band device-to-device communication in OFDMA cellular networks: A survey and challenges. *IEEE Commun. Surv. Tutor.* **2015**, *17*, 1885–1992. [CrossRef]
39. Rohde & Schwarz. *Device to Device Communication in LTE*; Rohde & Schwartz Whitepaper; Rohde & Schwarz USA, Inc.: Columbia, MD, USA, 2015.
40. 3rd Generation Partnership Project (3GPP). RP-170295 Study on Further Enhancements to LTE Device to Device, UE to Network Relays for IoT and Wearables. March 2017. Available online: http://www.3gpp.org/ftp/tsg_ran/TSG_RAN/TSGR_75/Docs/RP-170295.zip (accessed on 27 December 2017).
41. 3rd Generation Partnership Project (3GPP). TS 22.261 Service Requirements for the 5G System, V16.0.0. June 2017. Available online: http://www.3gpp.org/ftp//Specs/archive/22_series/22.261/22261-g00.zip (accessed on 27 December 2017).
42. 3rd Generation Partnership Project (3GPP). RP-161313 Report of 3GPP TSG RAN Meeting #71. June 2016. Available online: http://www.3gpp.org/ftp/tsg_ran/TSG_RAN/TSGR_72/Docs/RP-161313.zip (accessed on 27 December 2017).
43. Roessel, S.; Sesia, S. Cellular Internet-of-Things—Explained. In Proceedings of the IEEE Globecom Conference, Washington, DC, USA, 4–8 December 2016.
44. 3rd Generation Partnership Project (3GPP). RP-172119 Draft Report of 3GPP TSG RAN Meeting #77. December 2017. Available online: http://www.3gpp.org/ftp/TSG_RAN/TSG_RAN/TSGR_78/Docs/RP-172119.zip (accessed on 27 December 2017).
45. 3rd Generation Partnership Project (3GPP). TS22.186 Enhancement of 3GPP Support for V2X Scenarios, V.15.2.0. September 2017. Available online: http://www.3gpp.org/ftp//Specs/archive/22_series/22.186/22186-f20.zip (accessed on 27 December 2017).

46.   3rd Generation Partnership Project (3GPP). TR 22.819 Feasibility Study on Maritime Communication Services over 3GPP System, V0.3.0. August 2017. Available online: http://www.3gpp.org/ftp//Specs/archive/22_series/22.819/22819-030.zip (accessed on 27 December 2017).
47.   Capone, A.; D'Elia, S.; Filippini, I.; Redondi, A.E.C.; Zangani, M. Modeling energy consumption of mobile radio networks: An operator perspective. *IEEE Wirel. Commun.* **2017**, *24*, 120–126. [CrossRef]

future internet

MDPI

Article

Blockchain-Empowered Fair Computational Resource Sharing System in the D2D Network [†]

Zhen Hong *, Zehua Wang, Wei Cai and Victor C. M. Leung

Department of Electrical and Computer Engineering, The University of British Columbia,
Vancouver, BC V6T 1Z4, Canada; zwang@ece.ubc.ca (Z.W.); weicai@ece.ubc.ca (W.C.);
vleung@ece.ubc.ca (V.C.M.L.)
* Correspondence: hongz@ece.ubc.ca; Tel.: +1-604-353-5355
† Z. Hong, Z. Wang, W. Cai and V. C. M. Leung, "Connectivity-Aware Task Outsourcing and Scheduling in D2D Networks", 2017 26th International Conference on Computer Communication and Networks (ICCCN), Vancouver, BC, Canada, 31 July–3 August 2017; pp. 1–9.

Received: 8 October 2017; Accepted: 14 November 2017; Published: 17 November 2017

Abstract: Device-to-device (D2D) communication is becoming an increasingly important technology in future networks with the climbing demand for local services. For instance, resource sharing in the D2D network features ubiquitous availability, flexibility, low latency and low cost. However, these features also bring along challenges when building a satisfactory resource sharing system in the D2D network. Specifically, user mobility is one of the top concerns for designing a cooperative D2D computational resource sharing system since mutual communication may not be stably available due to user mobility. A previous endeavour has demonstrated and proven how connectivity can be incorporated into cooperative task scheduling among users in the D2D network to effectively lower average task execution time. There are doubts about whether this type of task scheduling scheme, though effective, presents fairness among users. In other words, it can be unfair for users who contribute many computational resources while receiving little when in need. In this paper, we propose a novel blockchain-based credit system that can be incorporated into the connectivity-aware task scheduling scheme to enforce fairness among users in the D2D network. Users' computational task cooperation will be recorded on the public blockchain ledger in the system as transactions, and each user's credit balance can be easily accessible from the ledger. A supernode at the base station is responsible for scheduling cooperative computational tasks based on user mobility and user credit balance. We investigated the performance of the credit system, and simulation results showed that with a minor sacrifice of average task execution time, the level of fairness can obtain a major enhancement.

Keywords: D2D communication; blockchain; fairness; connectivity-aware

---

1. Introduction

Advances in computing technology are transforming the way people execute computational tasks for daily applications like stock trading [1], gaming [2], etc. Usage of traditional desktop computers for large computational works has been expanded to various ways of computing such as cloud computing. For example, cloud gaming platforms PlayStation Now [3] and GameFly [4] execute most gaming computational tasks on the cloud, which frees gamers from having to update their computing devices frequently. Stock market investors are now able to manipulate stock trading on their mobile devices by offloading most computational tasks to the cloud [1].

In recent years, with the explosion of smart mobile devices and their capacities in terms of computing power, storage, data transmission efficiency, etc., the concept of fog computing [5] and D2D offloading has been facilitated to overcome high cloud service costs and mobility constraints.

Although it is widely adapted contemporarily with offloading of computational tasks to the cloud as the fog does not have as high a "density" (i.e., calculation and storage capacities), fog computing and D2D offloading prevent high carrier data transmission cost and cloud service costs, and its presence in users' vicinity can prevent high communication latency. Intermittent access to cellular data and non-seamless wireless coverage in the mobile environments are also discouraging factors for users to completely rely on the cloud. Faster and more responsive task cooperation and offloading in the D2D network becomes even more necessary in extreme situations like earthquake response.

The work in [6] shows that despite increasing usage of mobile devices in our daily lives, most of the computational power of these smart devices is still in the idle state and wasted, e.g., only email notification listeners and other low consumption applications run in the background for most of the time. If we can take advantage of the computational power of these idle devices together with their storage and data layover abilities, cost-effective task cooperation in D2D networks is highly feasible. Such a task cooperation and offloading context was first presented in Serendipity [7], a system that allows a mobile initiator to utilize computational resources available in other mobile systems in its surroundings to accelerate computing and save energy, whose performance is further analysed in [8] to see significant potential gain in both execution time and device energy. The authors of [9] proposed a mobile application that enables the cooperation of computationally-intensive applications by making use of computational powers of mobile devices in a nearby cloudlet.

While many previous works tried to exploit how idle computational power can be effectively utilized in D2D networks, the mobility aspects of users, especially the task cooperation scheduling in mobile environments, still remain open issues. Previous work in [10] illustrated a computational task cooperation system in the D2D network that provides users with significantly lowered task execution time without turning to cloud services that may introduce high monetary costs. However, this work does not consider the incentive for a user to share the computational resource of her/his device even though her/his device might be idle, neither is the fairness among users considered.

The work in [11] presented a reputation system incorporated with an ad hoc cloud gaming system. Without such a reputation system, unfairness will present as the players with higher network quality will be sacrificing significantly higher bandwidth that may lead to much higher monetary cost than those with lower network quality. In a D2D computation offloading system, similarly, unfairness may also result if users who contribute many computational resources are offered little, or even none, when in need. Therefore, it becomes important for us to build a reliable credit system on top of our computational resource sharing system to provide incentives for users to share their spare computational resources and enforce fairness while not affecting system effectiveness too much. Among various possible ways to implement a credit system for our computational resource sharing system, the recent upsurge of attention toward de-centralized blockchain technology has inspired us. Blockchain technology features de-centralized autonomy, anonymity, transparency, immutability, etc. [12], naturally meeting our system needs and becoming the choice as the basis of our credit system. In this work, we will be the first to propose a task outsourcing and scheduling scheme that is probabilistically based on the mobility of smart mobile device users in a D2D network, with a blockchain-based credit system to enforce fairness among users in the system.

The remaining parts of this work are organized as follows. Section 2 conducts a review on related works. Section 3 presents the system overview, and Section 4 models the proposed system. Section 5 illustrates the problem formulation of our proposed scheme, and Section 6 shows corresponding experimental evaluation results. Section 7 discusses the benefits and limitations of the proposed scheme. Section 8 concludes our work in this paper.

2. Related Work

2.1. Abundance of Spare Resources in D2D Networks

To relieve the burden of wireless cellular networks and the cloud, mobile data and computational traffic can be delivered through other means to the users (e.g., WiFi, D2D communications). This is known as mobile data and computation offloading. Several works have identified the benefits of WiFi data offloading [13–15]. The work in [13] showed that deferring the uploading tasks until WiFi access points are available can save the energy of smartphones. By jointly considering the power consumption and link capacity of wireless network interfaces, Ding et al. in [14] studied the criterion of downloading data from WiFi, as well as the WiFi access point selection problem.

However, mobile data traffic cannot always be offloaded to WiFi networks since the number of open-accessible WiFi access points is limited [14], just as the availability of affordable cloud computing services may be quite limited [5]. To fully exploit the benefits of data and computation offloading, mobile traffic and computational works can also cooperate in D2D networks. Specifically, mobile devices in close proximity can be connected via WiFi Direct [16], Bluetooth, etc., in a D2D manner for data and task cooperation between users. This is referred to as D2D data and computation offloading. The works in [7,8] explore task cooperation of mobile devices in the D2D network and showed that significant execution time and device energy can be saved. The authors of [17] presented a framework for opportunistic storage and processing in the mobile cloud. The work in [18] considers D2D technologies as candidates to deal with most local communications and time-sensitive computations in the near future. A D2D network should make use of Bluetooth, WiFi-Direct and other protocols to more efficiently provision services to applications such as video gaming and image processing.

It has been shown in [6] that the computational power of our smart mobile devices is idle and wasted for most of the time before these devices become outdated and replaced with newer models. The work in [19] presents that contemporary smart devices (mostly quad-core devices) use less than two cores on average in their non-idle states with the consideration of simultaneously running applications in the background, not to mention the computing power that these devices can provide in their idle states. It is generally true that building more data centres can provision more computational power for end users. However, a data centre needs to be built and maintained at a very high cost, which encourages us to exploit the task cooperation possibilities in D2D networks bearing users' mobility.

Considering the mobile nature of smart device users in ad hoc networks, Wang et al. in [20] proposed a metric, expected available duration (EAD), based on the mobility and similarities of users' interests in the D2D network. EAD indicates the statistically determined expected duration of each user's files of interest in the D2D network. With this metric, this work presents an optimization and performance promotion of a file sharing system in the D2D network to reduce the expensive data charge from cellular carriers and download more data from neighbours.

The work in [11] presented a reputation system incorporated with an ad hoc cloud gaming system that can reduce system players' overall bandwidth consumption while keeping fairness among them, without which players with higher network quality will be sacrificing significantly higher bandwidth. Similarly, if a user in our D2D computation offloading system can choose not to share spare computational resources while only receiving help from peers, it is unfair for those helpers contributing their computational resources. Consequently, we need to add a reliable credit system for our computational resource sharing system to enforce fairness among users, but not affecting system effectiveness too much. Multiple candidates are available for building a credit system, among which de-centralized blockchain technology seems to meet our system needs most.

2.2. Fairness and Blockchain

The authors in [21] presented a blockchain-based reputation system framework for joint cloud computing services, which evaluates the credibility of cloud service vendors in terms of service quality.

The blockchain-based information database stores vendor reputation values in a distributed manner and prevents the reputation values from being artificially tampered with, which benefits agnostic end users. The recent upsurge of attention toward de-centralized blockchain technology resulted because traditional credit systems like centralized banking and membership services are losing user confidence because users are agnostic and not truly in charge of their accounts. For example, if the cloud service vendor in [21] can easily tamper with it and increase its reputation value, the system is not trustworthy with respect to its customers. To build up a fair and trustworthy computational resource sharing system, blockchain technology naturally becomes the key cornerstone of our credit system. First, the blockchain needs to be maintained by mining (to be explained below), which can be performed by any of our system nodes. Second, the blockchain is available to all users, which is transparent and immutable so that users are in charge of their own accounts and transactions. Third, the transactions on the blockchain are anonymized, which provides user privacy, just to name a few. In this section, we describe the key concepts related to blockchain technology in general.

- Blockchain: Blockchain is a distributed data structure consisting of a chain of blocks. Blockchain works as a distributed database or a public ledger that keeps records of all transactions in the blockchain network. The transactions are time-stamped and listed into blocks where each block is identified by a unique cryptographic hash. Each block links to it previous block by referencing the hash value of the previous block, forming a chain of blocks and thus called a blockchain. A blockchain is maintained by a network of nodes, and every node records the same transactions. The blockchain is publicly accessible among the nodes in the blockchain network. Figure 1 illustrates the structure of a blockchain.

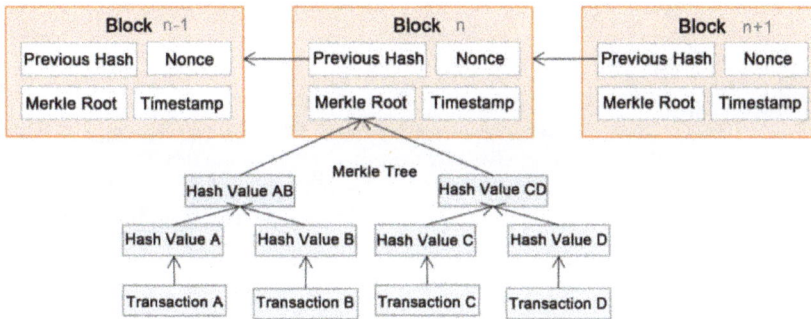

Figure 1. Typical blockchain structure.

- Blocks: The transactions in a blockchain network are bundled into blocks. These blocks are executed and maintained by all nodes in the network. A block consists of its hash, the hash of its previous block, a nonce that is used to avoid malicious nodes from flooding the network, a transaction list and a timestamp. In order to save storage space, the transaction list is typically stored in a Merkel root [22] format in each block. Only one of the conflicting transactions (e.g., transactions trying to double spend) will be taken as a part of the block. The blocks are added to the blockchain at regular intervals by miners.
- Transactions: A transaction is between two nodes in the blockchain network. Each transaction mainly includes the addresses of the sender and recipient, as well as a transaction value. In a valid transaction, the transaction value is transferred from the sender to the recipient. All transactions are signed by the sender's private key as a digital signature. Transactions are chosen and included in the blocks in the mining process. All transactions on a blockchain can be accessed by all participant nodes in the network.
- Mining: Transactions in a blockchain network are verified in a process called mining. Incentives, in the form of credit or crypto-currency, are provided to participating nodes to perform the mining

operations. Nodes participating in mining are called miners. A miner typically is required to select new transactions from a transaction pool, include them in a candidate new block and perform a mathematical computation to determine an appropriate nonce for the new block. This process of performing the mathematical computation is referred to as "proof of work" (PoW) [23], which is mainly used to prevent malicious nodes from arbitrarily adding new blocks to the blockchain or "flooding" the network. The first miner to come up with a valid nonce and thus a valid new block gets the block reward. Miners produce blocks that are then verified by other miners in the network for validity. Once a new winning block is selected, all other miners update to that new block. The longer the blockchain becomes, the harder for a malicious node to tamper with it. Therefore, mining is typically the key to keep data safety in blockchain applications. While mining is prevalent in contemporary blockchain applications, it is not necessary, and the discussion of this remains beyond the scope of this work.

3. System Overview

Our system consists of two major parts: the cooperative task scheduling to enhance effectiveness (e.g., average task execution time) among users and a blockchain-based credit system to provide fairness and incentives to users. Specifically, as the recent upsurge of interest in de-centralized blockchain technology suggests, traditional credit system like centralized banking and membership services are losing user confidence because users are not truly in charge of their accounts. The central power is able to modify user credit or create credit out of nothing, which can lead to user losses. Consequently, our credit system will be empowered by blockchain technology to enforce fairness and other benefits, e.g., autonomy and anonymity, among users, which effectively enhances user QoE in a fair manner.

3.1. Cooperative Task Scheduling and Roles of System Users

As shown in Figure 2, our D2D network consists of users with smart devices and a supernode at the base station (BS). Communications between user devices are through direct D2D links like Bluetooth or WiFi Direct, and communications between user devices and the supernode (e.g., reporting mobility and task information) are through a cellular link like 4G or LTE. D2D task cooperation is coordinated by the supernode bearing the mobility and task information among users in mind. In this paper, we assume that some necessary information related to a properly sliced task piece (including some overhead and necessary execution files, which are assumed to be of limited size not comparable to large multimedia content) will be sent from a requester to a helper, and the calculation result (which is even smaller) will be sent back to the requester once the helper has finished. The information exchanged between a user node and the supernode will be of a much more limited size, whose transmission time can also be negligible compared to the cooperative task execution time. More importantly, the D2D task cooperation is coordinated and assigned by the supernode at the base station, meaning that each helper is assigned specific time slots (to be elaborated in Section 5) and a corresponding amount of work to help each requester. Thus, we do not emphasize the difficulty of assigning dedicated in-band channels for the D2D communications in our system. Instead, we emphasize the difficulty of a user executing her/his own task in a timely manner, and since our system is not proposed for content sharing that is bandwidth significant, we assume that D2D communications between user nodes use dedicated in-band channels assigned by the supernode. Hence, mutual interference is not emphasized in our work.

At any moment, we may further divide system users into computational resource users and miners. Miners will write transactions into the main blockchain and grant credits for keeping our blockchain-based credit system safe. Computational resource users consist of requesters and helpers: requesters in an task period are devices in need of computational assistance, and helpers are devices that may offer help requesters. Each successful computational assistance will be recorded as a transaction and will be written into the blockchain. Therefore, a requester will need to pay the corresponding amount of credit to a helper after receiving the computational assistance from that helper.

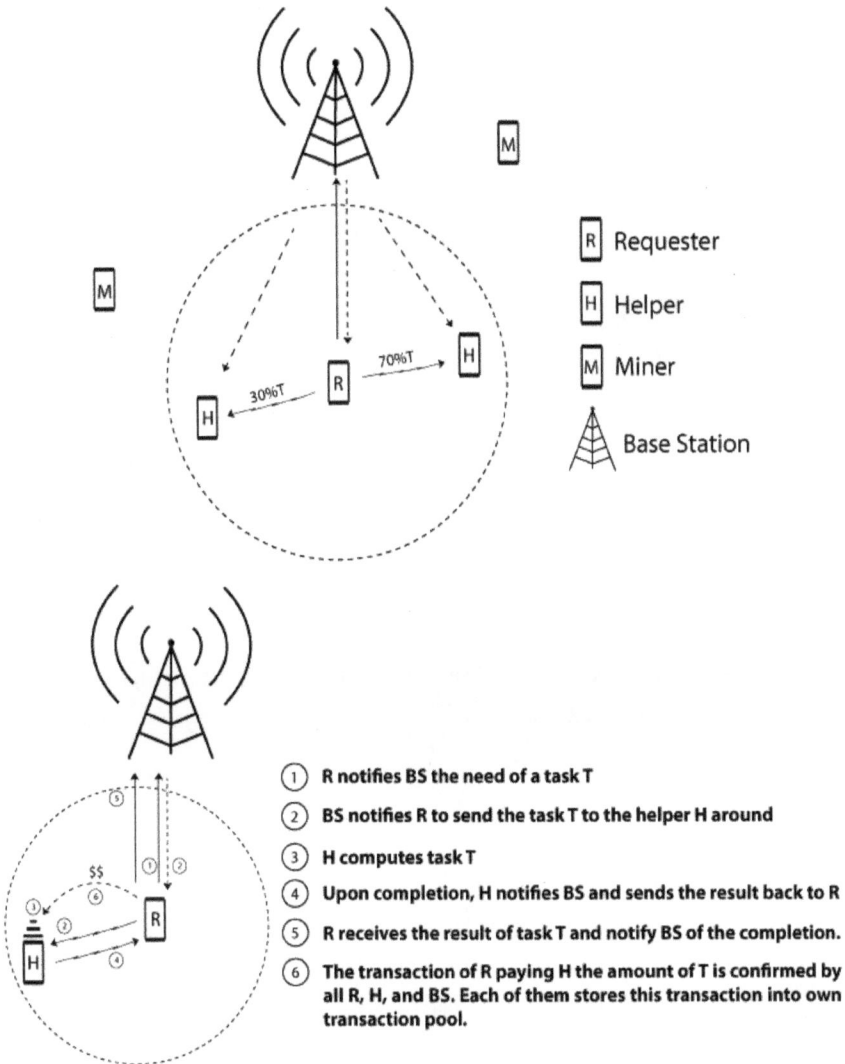

Figure 2. Task scheduling in the D2D network.

3.2. Supernode Coordination and Working Process

In our system, the supernode will not only assign task assignments to devices in the D2D network, it will also work as a coordinator between a requester and worker pair by acknowledging their cooperation work. As shown in Figure 2, the step-by-step working process is as follows: (1) A requester notifies the supernode at the BS about the need for a cooperative task T; (2) After calculation and analysis of the system conditions, the supernode will assign, say, 30% of cooperative task T to one helper and 70% of T to another helper around the requester. The supernode will notify the requester and each related helper about the cooperation assignment information. The requester then will send corresponding task portions to each assigned helper; (3) Each helper executes the task portion on the device; (4) Upon completion, each helper will notify the supernode and send back the result to the requester; (5) When the requester gets back the computational result from a helper, she/he will notify the supernode about the successful reception of the result; (6) A transaction of the requester paying each related helper is confirmed by all three: the requester, the helper and the supernode. All three of them will store this transaction into their own transaction pool, waiting for a miner to put this transaction into the blockchain.

3.3. Transaction Pool and Mining

After cooperation is performed between a requester and a helper, both of them will have an identical transaction generated. They will store this transaction into their own transaction pool on their own device and also broadcast to nearby peers. Each peer will then store the transaction into her/his own transaction pool upon reception of the broadcast transaction. Note that the supernode has the information of all transactions in the D2D network, so the supernode holds the publicly assessable full transaction pool for all users in the network. The storage space of the transactions is negligible. Figure 3 is a typical part of the blockchain of our system. Each block is identified with a 256-bit unique hash value and links to its previous block. Each block contains transactions, also identified with a 256-bit unique hash value, that contain information about the cooperative work in the D2D network. For example, Transaction 1 in the left block is recording that User 1 paid 2000 credits to User 2 for receiving the corresponding amount of helper work from User 2. As a safety feature, each transaction needs to point to the source of the income of the credit, as a proof of enough credit. For instance, Transactions 1 and 2 at the right block both point to Transaction 1 in the left block since this is an indication that User 2 does have enough credit to pay the total of 1600 credits in the right block. Similarly, Transaction 3 at the right block also points to Transaction 2 in the left block, indicating that User 4 has enough credit. Note that for demonstration purposes, users are labelled as U_1, U_2, etc., in the figure. In fact, these users are actually represented as 256-bit unique digital addresses to provide anonymity. Users are also able to change their addresses to further enhance anonymity.

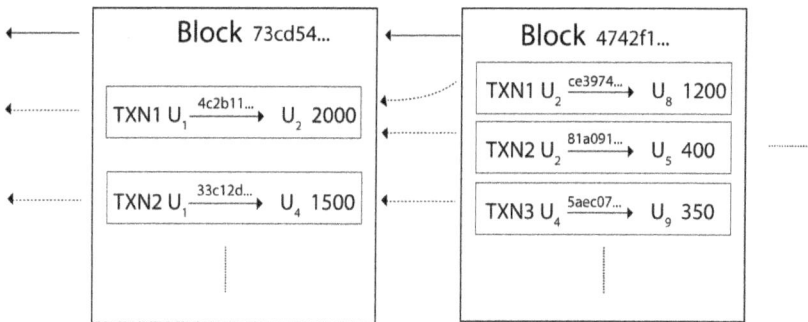

Figure 3. Typical part of the blockchain of our system.

A user will switch between a helper and a miner according to whether task cooperation is needed in the surrounding area. When a device is assigned work to help a requester in the D2D network, the device will switch to helper mode; otherwise, i.e., when no task is assigned to assist a helper in the current period, the device will switch to miner mode in search for a mining possibility. The debate between how secure the mining algorithm is compared to those used in Bitcoin or Ethereum remains beyond the scope of this work.

4. System Modelling

In this section, the basic system settings, connectivity model, dynamic program slicing, task cooperation scheduling background and the credit balance system that is blockchain based are illustrated.

4.1. Basic System Settings

Apart from the set of all miners in the D2D network, there is a computational resource user set U consisting of u users in our system, where in each task period p, u users are requesters and the rest $h = u - p$ are helpers. We divide the user set U into two member sets, namely requester set P and helper set H. There are p requesters in P and h helpers in H, where $p \geq 1$ and $h \geq 1$. Each requester is denoted as $p_i \in P$, and each helper is denoted as $h_j \in H$, where $i \in \{1, 2, ..., p\}, j \in \{1, 2, ..., h\}$. The smart device of a requester p_i or helper h_j is subject to a D2D communication range r_i^p or r_j^h, respectively, above which D2D direct link connection is not possible. c_i^p or c_j^h is used to denote the available computational power of a requester or helper, indicating how fast or how much computation the smart device is able to handle per second for our cooperative scheme. Typically, this type of computational power is represented by how many clock cycles the device can run per second, 2.6 GHz for example. At any task period $\Psi \geq 0$, each requester p_i initializes a task of complexity $T_{i,\Psi}$ in clock cycles (indicating how many clock cycles need to be run to get the result of the task) with its maximum wait time $t_{i,\Psi}$ in seconds (indicating the maximum time p_i will wait for the result until he/she has to do the assigned uncompleted task slices by himself/herself).

4.2. Connectivity Model

Assuming the connection between a requester p_i and a helper h_j to be symmetric, we denote the random variable $B_{i,j}(\tau) = 1$ (or $B_{i,j}(\tau) = 0$) to represent that p_i and h_j are connected (or disconnected) at time $\tau \geq 0$. Moreover, let random variable $S_{i,j}^1$ denote the sojourn time that p_i and h_j are in the connected state and $S_{i,j}^0$ denote that in the disconnected state. We consider that both $S_{i,j}^1$ and $S_{i,j}^0$ follow the exponential distribution with parameters $\lambda_{i,j}$ and $\mu_{i,j}$, respectively. Therefore, we have the cumulative distribution functions (CDF) of $S_{i,j}^1$ and $S_{i,j}^0$ given by:

$$\Pr(S_{i,j}^1 \leq \tau) = 1 - e^{\mu_{i,j}\tau}, \tag{1}$$

and:

$$\Pr(S_{i,j}^0 \leq \tau) = 1 - e^{\lambda_{i,j}\tau}. \tag{2}$$

We represent the continuous time Markov chain (CTMC) model with two states illustrated in Figure 4 and let $P_{i,j}(\tau)$ denote the 2×2 matrix with entries $p_{i,j}^{xy}(\tau) = \Pr(B_{i,j}(\tau) = y | B_{i,j}(0) = x)$, where $x, y \in \{0, 1\}$. Referring to [24], we have the solution of $P_{i,j}(\tau)$ given by:

$$P_{i,j}(\tau) = \begin{pmatrix} \frac{\lambda_{i,j}}{\psi} + \frac{\mu_{i,j}}{\psi}\kappa & \frac{\mu_{i,j}}{\psi} - \frac{\mu_{i,j}}{\psi}\kappa \\ \frac{\lambda_{i,j}}{\psi} - \frac{\lambda_{i,j}}{\psi}\kappa & \frac{\mu_{i,j}}{\psi} + \frac{\lambda_{i,j}}{\psi}\kappa \end{pmatrix}, \tag{3}$$

where $\kappa = e^{-(\mu_{i,j}+\lambda_{i,j})\tau}$ and $\psi = \mu_{i,j} + \lambda_{i,j}$.

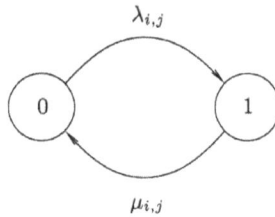

Figure 4. Continuous-time Markov chain for the connected-disconnected transition between p_i and h_j. Let 1/0 represent the state in which p_i and h_j are connect/disconnected. The transition rates from zero to one and from one to zero are given by $\lambda_{i,j}$ and $\mu_{i,j}$, respectively.

The parameters $\mu_{i,j}$ and $\lambda_{i,j}$ used in (3) for p_i and h_j can be obtained by maximum likelihood estimation (MLE) on each of them. Specifically, without loss of generality, consider that p_i and h_j were disconnected initially, and the connectivity between p_i and h_j has changed m times before the current time τ. Therefore, p_i and h_j have recorded a vector of time $\vec{\tau}_{i,j} = (\tau_1, ..., \tau_m) \in R_+^m$, where each element $\tau_{i,j}^z < \tau(z = 1, ..., m)$ represents the time when the connectivity between p_i and h_j changed. Assume p_i and h_j are currently connected (then, m must be an odd number, and the case that p_i and h_j are currently disconnected can be analysed via following approach similarly); $\mu_{i,j}$ and $\lambda_{i,j}$ estimated by MLE up to current time τ are given by:

$$\hat{\mu}_{i,j}^t = \frac{m-1}{2\sum_{z=1}^{\frac{m-1}{2}} \left(t_{i,j}^{2z} - t_{i,j}^{2z-1}\right)}, \tag{4}$$

and:

$$\hat{\lambda}_{i,j}^t = \frac{m-1}{2\sum_{z=1}^{\frac{m-1}{2}} \left(t_{i,j}^{2z+1} - t_{i,j}^{2z}\right)}. \tag{5}$$

Because the connections between p_i and h_j are assumed symmetric, the same results are obtained on p_i and h_j. For the people who study or work together, $\vec{\tau}_{i,j}$ is kept being recorded by both p_i and h_j as the system time increases. According to (4) and (5), $\hat{\mu}_{i,j}^t$ and $\hat{\lambda}_{i,j}^t$ will converge. We denote $\hat{\mu}_{i,j} = \lim_{t\to\infty} \hat{\mu}_{i,j}^t$ and $\hat{\lambda}_{i,j} = \lim_{t\to\infty} \hat{\lambda}_{i,j}^t$, which are the MLE of $\hat{\mu}_{i,j}$ and $\hat{\lambda}_{i,j}$, respectively. Given the connection station $B_{i,j}(\tau)$ between p_i and h_j at time τ, the probability that they are connected at future time $\tau' \geq \tau$ is given by:

$$\Pr(B_{i,j}(\tau') = 1 | B_{i,j}(\tau)) = \begin{cases} \frac{\lambda_{i,j} - \lambda_{i,j} e^{-(\lambda_{i,j} + \mu_{i,j})(t'-t)}}{\lambda_{i,j} + \mu_{i,j}}, & B_{i,j}(\tau) = 0, \\ \frac{\lambda_{i,j} + \mu_{i,j} e^{-(\lambda_{i,j} + \mu_{i,j})(t'-t)}}{\lambda_{i,j} + \mu_{i,j}}, & B_{i,j}(\tau) = 1. \end{cases} \tag{6}$$

4.3. Dynamic Program Slicing

In general, computational tasks cannot be arbitrarily sliced into different parts. However, many dynamic program slicing techniques are facilitating our need for the distribution of tasks [25]. For example, MapReduce [26] allows Google to slice and run an average of one hundred thousand MapReduce jobs every day from 2004–2008. Without the availability of dynamic program slicing, a large load of tasks may not be sent back to the requester in a timely manner, and this increases the risk of the helper being out of the device communication range of the requester on completion of the task execution. Therefore, we adopt the assumption that tasks can be sliced in an arbitrary manner in our system.

4.4. Task Cooperation Scheduling

In our system, we assume that a computing unit, which we refer to as a supernode, with enough capacity to perform task cooperation scheduling for all devices is available at the cellular BS covering the D2D network. At the beginning of each task period, the supernode collects task and connectivity information sent wirelessly from the devices and computes the task scheduling for requesters and helpers in the D2D network based on the probability of connection among them according to (6). Note that the task and connectivity information sent to the supernode is very limited in size and transmission time, which are assumed to be negligible for simplicity. Each task period is divided into discrete time slots for scheduling to ensure accuracy and latency and to lower the chance of losing computation results due to changes in connectivity. This computation will be based on an effective light-weight algorithm elaborated in Section 5.5.

4.5. Blockchain-Empowered Credit System

In contrast to a Bitcoin system, where later, the user is discouraged from joining due to the significantly increased difficulty to obtain a new coin, our system offers the same initial credit, ω, to any new user to the system. Users need to pay credits from their own balance to get help from peers and will earn credits after helping peers on computational tasks. At the beginning of each task period $\Psi \in \{1, 2, 3, ...\}$, the credit balance B_k^Ψ of each user u_k is obtained by the supernode by referring to the blockchain. For each user u_k in period Ψ, we denote the amount of help received by peers as R_k^Ψ, the amount of work contributed to peers as H_k^Ψ and the block reward as η_k^Ψ. For simplicity, we limit a user to be either a requester, a helper or a miner within a given task period Ψ. In our blockchain-empowered credit system, u_k needs to pay αR_k^Ψ from and is rewarded βH_k^Ψ to the balance B_k^Ψ; therefore,

$$B_k^{\Psi+1} = \begin{cases} B_k^\Psi - \alpha R_k^\Psi + \beta H_k^\Psi + \eta_k^\Psi, & \Psi \geq 1, \\ \omega, & \Psi = 0. \end{cases} \tag{7}$$

5. Problem Formulation

As mentioned in the previous section, we assume that a supernode with enough capacity is present at the BS covering the D2D network of our interest. At the beginning of each task period, the supernode will, with the knowledge of all devices and tasks in the D2D network, calculate the probability distribution of the connectivities between devices and assign computational tasks to each helper or requester device accordingly. The task assignment also takes into account users' credit balance when our blockchain-based credit system is adopted. Connectivity awareness, computational task assignment, selfishness avoidance and response delay optimization are mathematically formulated in this section.

5.1. Connectivity Awareness

To analyse the connectivity between requesters and helpers for more accurate and cost-effective computation assignment, we first define a $p \times (h+1)$ probability matrix R_t at each time slot $t \in \{0, 1, ..., \frac{T}{\Delta t}\}$. For a given t, the element $R_{i,j,t}$ is the probability of connection between p_i and h_j if $j \in \{1, 2, ..., h\}$ and $R_{i,(h+1),t}$ indicating the probability of connection between p_i and herself/himself. Obviously, $R_{i,(h+1),t} = 1 \; \forall \, i, t$. At the start of each task period, i.e., $t = 0$, we randomly generate $R_{i,j,0} \; \forall \, i \in \{1, 2, ..., p\}, j \in \{1, 2, ..., h\}$ based on the pre-defined initial connection probability. Thereafter, according to Equation (6), we generate $R_{i,j,t} \; \forall t \in \{1, 2, ..., \frac{T}{\Delta t}\}$ for each p_i-h_j pair.

At any given time slot $t \in \{1, 2, ..., \frac{T}{\Delta t}\}$, we define a p-element vector $\vec{I}_t$ with its element $\vec{I}_{i,t}$ representing the amount of self-computing computational task assigned to p_i and a $p \times h$ matrix J_t with element $J_{i,j,t}$ representing the amount of assisting computational task assigned to h_j for p_i. By concatenating $J_{i,j,t}$ and $\vec{I}_{i,t}$, we get a $p \times (h+1)$ computation assignment matrix $M_t = [J_{i,j,t} \; \vec{I}_{i,t}]$ at time slot t. Joining all M_t where $t \in \{1, 2, ..., \frac{T}{\Delta t}\}$, we get a $p \times (h+1) \times \frac{T}{\Delta t}$ three-dimensional system computation assignment matrix, M, containing the computation assignment in a task period $\tau \in [0, T]$. Consequently, $M_{i,j,t}$ where $j \in \{1, 2, ... h\}$ is the amount of computational task assigned to h_j for p_i and $M_{i,h+1,t}$ is the amount of self-computing computational task assigned to p_i at time slot $\tau = t$.

Note that the element-wise product between R and M,

$$M^{exp} = M \odot R, \tag{8}$$

is the matrix indicating the expected amount of computational task done and the result sent back to requesters. For example, $M_{i,j,t}^{exp} = M_{i,j,t} \cdot R_{i,j,t}$ is the expected amount of assisting task done by h_j and sent back to p_i at time slot $\tau = t$.

5.2. Computation Assignment and Maximum Wait Time

At the start of each task period, all requesters will specify to the supernode at the BS the amount of computation, in clock cycles, required for the coming task period. We represent these tasks with a p-element vector $\vec{\gamma}$ with γ_i corresponding to the total amount of task required by p_i. Meanwhile, for an ensured QoE, p_i is also subject to a maximum wait time in each task period for the result of the computational task. We use another p-element vector $\vec{\phi}$ with ϕ_i corresponding to the maximum wait time for p_i in seconds. Apparently, $\phi_i \leq T, \forall i$. Therefore, the computation assignment needs to ensure that a task is expected to be completed before the maximum wait time for all requesters, that is:

$$\gamma_i \leq \sum_{t=1}^{\phi_i/t} \sum_{j=1}^{(h+1)} M_{i,j,t}^{exp}, \forall i \in \{1, 2, ..., p\} \tag{9}$$

5.3. Computation Capacity of Mobile Devices

Each mobile device is subject to a computation capacity denoted as c_i^p for p_i's mobile device and c_j^h for h_j's mobile device where $i \in \{1, 2, ..., p\}, j \in \{1, 2, ..., h\}$. When talking about the computation capacity of a device, one typically will refer to its CPU. CPU processing capacity is typically referred to in terms of megahertz (MHz) or gigahertz (GHz). Professionals talk about clock speed, which is the standard ability of the CPU to cycle through its operations over time. Therefore, a 1-GHz CPU is able to tick its clock around one billion times per second, which in turn can perform more complicated computational tasks. Without loss of generality, we regulate these computational power values in clock cycles to a scale of 0–100, i.e., $0 \leq c_i^p \leq 100$ and $0 \leq c_j^h \leq 100 \ \forall \ i, j$, for simplicity. Each entry of the computation assignment matrix, $M_{i,j,t}$, refers to the number of clock cycles required to perform the corresponding task section. For example, $M_{3,4,0} = 1000$ means that at $t = 0$, h_4 is assigned to help p_3 for 1000 clock cycles worth of computational task. If $c_4^h = 50$ Hz, then it takes h_4 $\frac{M_{3,4,0}}{c_j^h} = \frac{1000}{50 \text{ Hz}} = 20$ s to perform the task. Therefore, each device is subject to an amount of computational task in clock cycles at each time slot as a higher limit, that is:

$$\sum_{i=1}^{p} M_{i,j,t} \leq c_j^h, \ \forall j \in \{1, 2, ..., h\}, t \in \{1, 2, ..., \frac{T}{\Delta t}\} \tag{10}$$

for all helper devices, and:

$$M_{i,h+1,t} \leq c_i^p, \ \forall i \in \{1, 2, ..., p\}, t \in \{1, 2, ..., \frac{T}{\Delta t}\} \tag{11}$$

for all requester devices as the $(h+1)$th column of the computation assignment matrix is representing the amount of self-computing tasks.

5.4. Selfishness Avoidance

As derived in Section 4.5, the balance of each user is updated as represented in (7). When computing cooperative task assignment for the D2D network at each task period, the supernode needs to make sure that each requester has enough balance for the task period according to the task assignment matrix M in that period Ψ. Since the exact amount of R_k^{Ψ} and H_k^{Ψ} for user u_k is not known at the beginning of task period Ψ, the supernode ideally needs to make sure that:

$$\Pr(B_k^{\Psi} - \alpha R_k^{\Psi} + \beta H_k^{\Psi} + \eta_k^{\Psi} \geq 0) \geq \xi, \forall k \in \{1, 2, ..., u\}, \Psi \geq 1. \tag{12}$$

where $\xi \to 1$. However, though the exact amount of R_k^{Ψ} and H_k^{Ψ} for user u_k is not known at the beginning of task period Ψ, their expected value, namely $R_k^{\Psi,exp}$ and $H_k^{\Psi,exp}$, can easily be obtained in advance from M^{exp} in that period Ψ:

$$R_k^{\Psi,exp} = \sum_{t=1}^{\phi_k/t} \sum_{j=1}^{h} M_{k,j,t}^{\Psi,exp}, \tag{13}$$

and:

$$H_k^{\Psi,exp} = \sum_{i=1}^{p} \sum_{t=1}^{\phi_i/t} M_{i,k,t}^{\Psi,exp}. \tag{14}$$

For simplicity, we will relax the constraint (12) to the following:

$$B_k^{\Psi} - \alpha R_k^{\Psi,exp} + \beta H_k^{\Psi,exp} + \eta_k^{\Psi} \geq 0, \forall k \in \{1, 2, ..., u\}, \Psi \geq 1. \tag{15}$$

This way, user connectivity has been taken into account, and more cooperation is expected to be done, while leaving the possibility that a user's balance becomes lower than zero after task period Ψ such that the user will need to earn back enough credit before asking for more help.

5.5. Response Delay Optimization

In our work, we emphasize the importance of low task execution time towards a requester's QoE. In each task period, there is $n = \frac{T}{\Delta t}$ time slots, and we define the expected completion time, t_i^{exp}, for each requester p_i as follows:

$$t_i^{exp} = \arg\min_t \sum_{\tau=1}^{t} \sum_{j=1}^{(h+1)} M_{i,j,t}^{exp} \geq \gamma_i, \tag{16}$$

where $i \in \{1, 2, ..., p\}, j \in \{1, 2, ..., h\}, t \in \{1, 2, ..., \frac{T}{\Delta t}\}$. Therefore, our optimization problem becomes:

$$\text{Minimize:} \quad \sum_{i=1}^{p} t_i^{exp} \tag{17}$$

$$\text{Subject to:} \quad (7) - (11) \ (13) - (16).$$

Calculation of the optimal computation assignment matrix M is similar to the famous knapsack problem that is NP-hard [27]. Here, the computational powers of helper devices are like knapsacks with different sizes, and the computational tasks from requesters are like items with different sizes. Solving for the optimal solution for the task scheduling resembles solving for an optimal solution for the knapsack problem: it is NP-hard. For effectiveness, especially considering the trend of an increasingly massive number of smart mobile devices in D2D networks, we proposed a light-weight heuristic algorithm to efficiently find the sub-optimal solution for the computation assignment matrix M illustrated in Algorithm 1. Note that we make substantial use of the linprog function [28] in MATLAB for calculating the computation assignment matrix M by transforming M element-wise into a one-dimensional unknown vector $\vec{x}$ with $p \cdot (h+1) \cdot \frac{T}{\Delta t}$ entries from elements in M. According to linprog [28], A, b in Algorithm 1 represent the inequality constrains for $\vec{x}$, and A_{eq}, b_{eq} represent the equality constrains for $\vec{x}$. The function get_linprog_parameter in Algorithm 1 is transforming the constraint functions in (17) from a matrix from to a one-dimensional vector form, which is basically a simple reshaping of the matrix. There is no upper bound for $\vec{x}$, and the lower limit for elements of $\vec{x}$ is zero. The maximum wait time of a requester is the longest time the requester can wait before she/he needs to compute the task result herself/himself. However, it is possible that the task can be completed much sooner than the maximum wait time. Each iteration of Algorithm 1 tries to find lower feasible task assignment solutions by halving (up to an integer value) a randomly-selected requester's maximum wait time. Since the duration of each task period is limited, the maximum wait time is up to the length of a task period. Therefore, the complexity of Algorithm 1 is subject to $ln(p) \cdot O(linprog)$ where p is the number of requesters at the task period and $O(linprog)$ represents the complexity of MATLAB's linprog function [28]. Unfortunately, MATLAB claims improvement on efficiency of linprog over time, but releases no detail about the complexity. Yang in [29] claims that his algorithm on top of linprog may achieve polynomial complexity with the best known complexity bound on linear programming problems.

Algorithm 1: The algorithm to obtain $\vec{\phi}$, which corresponds to a sub-optimal solution for (17).

> **Input:** Maximum wait vector $\vec{\phi}$
> **Output:** Modified maximum wait vector $\vec{\phi}$ that corresponds to a sub-optimal solution for (17)
> 1 **function** heuristicMaxWait($\vec{\phi}$)
> 2 $\quad [A, b, A_{eq}, b_{eq}] := \text{get\_linprog\_parameter}(\vec{\phi})$
> 3 $\quad [x, feasible] := \text{linprog}(\vec{0}, A, b, A_{eq}, b_{eq}, \vec{0})$
> 4 $\quad$ **if** $feasible$ **then**
> 5 $\qquad C := \{1, 2, ..., p\}$
> 6 $\qquad$ **while** $C \neq \emptyset$ **do**
> 7 $\qquad\quad$ **for** random $i \in C$ **do**
> 8 $\qquad\qquad \iota := \vec{\phi}_i$
> 9 $\qquad\qquad \vec{\phi}_i := \left\lfloor \frac{\vec{\phi}_i}{2} \right\rfloor$
> 10 $\qquad\qquad [A, b, A_{eq}, b_{eq}] := \text{get\_linprog\_parameter}(\vec{\phi})$
> 11 $\qquad\qquad [x, feasible] := \text{linprog}(\vec{0}, A, b, A_{eq}, b_{eq}, \vec{0})$
> 12 $\qquad\qquad$ **if** $!feasible$ **then**
> 13 $\qquad\qquad\quad \vec{\phi}_i := \iota$
> 14 $\qquad\qquad\quad C := C \backslash \{i\}$
> 15 $\quad$ **return** $\vec{\phi}$

---

6. Experiment

In a D2D network, a variety of factors may affect the performance of our cooperative network with the credit system. In this section, we will examine the following effects:

- Effect of initial credit: we vary the initial credit provided to each user to see how our system will be affected.
- Effect of mean maximum wait time: we vary the mean maximum wait time during the random generation to see how the performance will be affected.
- Effect of mean task size: we vary the mean task size of each requester in a task period during the random generation to see how the performance will be affected.
- Effect of time elapsed: we run the simulation on multiple task periods and see how system performance changes over time.

In our simulation, we compare the performance of the computational resource sharing system in four different cases:

- Greedy D2D task cooperation without our credit system: Without connectivity awareness, each helper device will equally contribute its available computing power to all connecting requesters at the beginning of a task period. For example, helper h_5 will assign $\frac{c_5^h}{3}$ computing power to each of p_1, p_3, p_4 for the current task period if and only if p_1, p_3, p_4 are the only requesters in connection with h_5 at time $\tau = 0$.
- Greedy D2D task cooperation with our credit system: This is very similar to the above case, except that our blockchain-based credit system is added in to enforce fairness. Therefore, the supernode at BS will check on requester balances before task assignment, ensuring that the assistance expected to be received by a requester will not exceed her.his available balance in that task period.
- Connectivity-aware task scheduling without our credit system: At the start of each task period, the supernode at BS will perform task scheduling calculation according to Algorithm 1 without our blockchain-based credit system.
- Connectivity-aware task scheduling with our credit system: This is very similar to the above case, except that our blockchain-based credit system is added in to enforce fairness, as illustrated in (15).

Apart from being a reasonable incentive for users to provide help and gain credit for future needs, the blockchain credit also provides selfishness avoidance to our system. That is, it prevents certain users who only want to get help from peers, but not contribute to other users' need. We define hereof the level of selfishness, LoS, to reflect whether users have been contributing relatively equally over task periods $\Psi \in \{1, 2, 3, ..., \chi\}$:

$$LoS = \frac{1}{u} \sum_{k=1}^{u} \left(\sum_{\Psi=1}^{\chi} R_k^{\Psi} - \sum_{\Psi=1}^{\chi} H_k^{\Psi} - \omega \right)^2 \tag{18}$$

In the following experimental illustrations, we show the comparison of LoS in a normalized way: with respect to the greedy D2D task cooperation without our credit system case since LoS for this case is much larger than the other three cases and it appears to be stable over the changing factors. For simplicity, the mining process is simplified to a random selection of idle users to be miners in the network. To validate the performance of our proposed system, we set up the following experiment. Default simulation parameters are illustrated in Table 1, where the uniform distribution between two values a, b is denoted as $U[a, b]$. We used three real-world traces "Intel" (Trace 1), "Cambridge" (Trace 2) and "Infocom" (Trace 3) in the Cambridge/Haggle dataset in [30] for our simulations. Traces 1–3 were recorded by 8, 12 and 41 mobile iMotes using Bluetooth with a 30-metre radio range, respectively. Although these iMotes were not smartphones or tablets, the connection states recorded in these traces can be used to reproduce the dynamic topology for mobile users. The interval of each iMote sending a beacon (i.e., hello message) is 120 ± 12 s.

The connectivity between mobile users is assumed to be symmetric in our work. However, the connect and disconnect events in traces were recorded by each iMote individually. Thus, we consider that a pair of iMotes was connected (or disconnected) as long as one of them detected a connect (or disconnect) event. In the real-world traces, an iMote has recorded a connect event with a zero contact duration when it was connected with another iMote for a short period of time such that the iMote failed to receive two or more consecutive beacons. Thus, for a record with the zero contact duration, we assume that the actual contact duration is uniformly distributed on [0 s, 120 s]. We concatenate the contact and inter-contact durations recorded by each pair of iMotes in a chronological order to reproduce the connect and disconnect events for both of them. We then run trace-driven simulations with the D2D topologies reproduced by all iMote pairs in each trace.

Table 1. Default simulation parameters.

Number of users u	10
Mean number of requester $\bar{p}$	$0.4u$
Mean number of miners	$0.1u$
Tasks period T	60 s
Size of time slots Δt	5 s
Total number of periods χ	30
Initial credit ω	5000
Block reward credit per task period η	50
α	1
β	1
Maximum computation capacity c^{max}	100
Minimum computation capacity c^{min}	40
Computation capacity c_i^p or c_j^h	$U[c^{min}, c^{max}]$
Mean task size per second σ	30 ($U[15, 45]$)
Mean maximum wait time	40 s ($U[20$ s, 60 s])
$\lambda_{i,j}$	$U[10^{-5}, 10^{-3}]$
$\mu_{i,j}$	$U[10^{-3}, 10^{-2}]$
Initial connection probability at $\tau = 0$	50%

6.1. Effect of Initial Credit

The choice of initial credit is a rather significant factor in our system. The initial credit is an indication of how much helper work can be received before a requester has to help others or perform mining to get system credits. If the initial credit is set too low, users are discouraged from performing D2D cooperation. In the extreme case in Figure 5, where initial credit is set to zero, the users cannot perform any D2D cooperation, resulting in a self-computing performance with a high average task execution time and no selfishness.

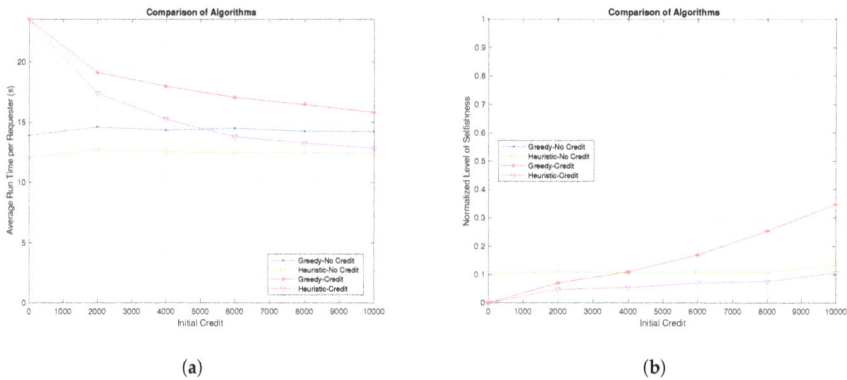

Figure 5. Effect of initial credit. (**a**) Effect on average task execution time; (**b**) effect on the level of selfishness.

As initial credit increases, users are encouraged to perform D2D cooperation, leading to a fast drop in average task execution time in the two cases with our credit system. Particularly, the average task execution time of the heuristic case with our credit system drops to <5% more than that without our credit system, while remaining 20% less selfish. This is an indication that with a proper selection of initial credit, a little sacrifice on average task execution time can be exchanged for a much higher level of fairness. The greedy-credit case has much lower *LoS* than that without the credit system, but its performance is much worse than the heuristic-credit case, implying the importance of our heuristic algorithm.

6.2. Effect of Mean Maximum Wait Time

During a task cooperation in the D2D network, the requester typically wants the task to be done in a timely manner. For example, if a requester wishes to perform a neural network-based stock index prediction task [31] that predicts the price of a stock in 30 s, this requester will want to get the computation result within 30 s (could be 20 s, 25 s, etc., depending on the specific application logic and handling behind).

As shown in Figure 6a, two heuristic cases started with a decrease in average runtime from mean maximum wait time changing from 15 s to 20 s. This may be due to the fact that when the mean maximum wait time is too low, the probability of finding a feasible solution in the cases with our heuristic algorithm is much lower. As mean maximum wait time continues to increase thereafter, the average task execution time increases in all four cases: the enlarged solution space is increasing the difficulty of our heuristic algorithm in finding the optimal solution; and in the greedy cases, the helpers cannot focus on helping fewer requesters since most requesters have similarly high mean maximum wait time. Particularly, the performance on the average task execution time of the greedy-credit case deteriorates approximately 36% from 14 s–19 s, while that for our heuristic case with our credit system deteriorates only 9% for the same change. This further proves the necessity of our heuristic algorithm. However, the normalized level of selfishness decreases with respect to the greedy-non credit case, with the two heuristic cases scaling down faster. Comparing the heuristic cases with and without our credit system, the sacrifice of around 10–15% of average task execution can bring us at least 40% less selfishness at a 50-s mean maximum wait time and almost 100% less at a 15-s mean maximum wait time.

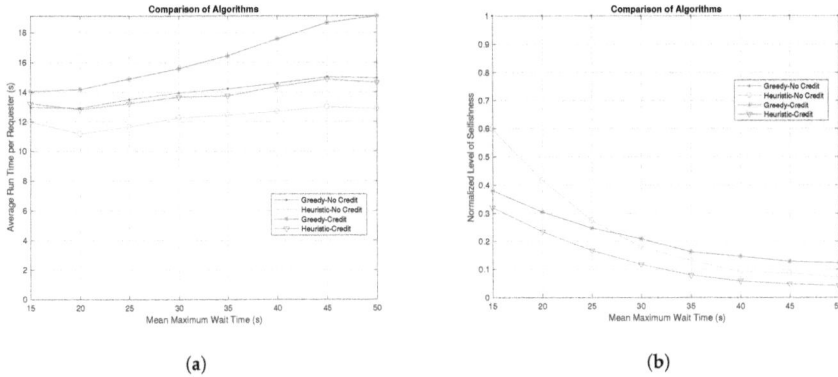

(a)   (b)

Figure 6. Effect of mean maximum wait time. (**a**) Effect on average task execution time; (**b**) effect on the level of selfishness.

6.3. Effect of Mean Task Size

The generation of task sizes in a period T is a uniform distribution $U[0.5\sigma T, 1.5\sigma T]$. Note that it is possible that the mean task size for a requester within a period is over the computing capacity of the requester device itself. Therefore, a D2D cooperation is necessary if the task needs to be done within the maximum wait time.

As illustrated in Figure 7a, the average task execution time increases with respect to the mean task size almost directly proportionally. The performance of our heuristic cases starts at a very close performance at the beginning when the mean task size is small; a requester does not need too much assistance work from helpers, leading to a relatively lower level of selfishness and average task execution time. When the mean task demand from requesters increases, the level of selfishness in the heuristic case without our credit system increases much faster than that with our credit system. When the normalized mean task size is 60, though the heuristic case without our credit system is 15%

better on the average task execution time, it is yet more than 250% higher in the level of selfishness. This shows how important it is to use our credit system to enforce fairness among users.

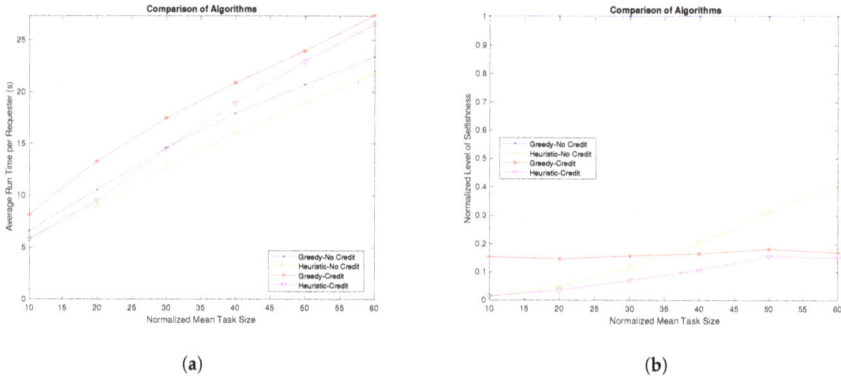

Figure 7. Effect of mean task size. (**a**) Effect on average task execution time; (**b**) effect on the level of selfishness.

6.4. Effect of Time Elapsed

Here, time elapsed is represented in the unit of the number of task periods elapsed. Effect of time elapsed generally gives an idea of how the performance of the four cases will stabilize over time.

As shown in Figure 8, the performance of our credit system starts to stabilize beyond the point of the 50th period. As the task period goes on, our blockchain-empowered credit system maintains a good level of selfishness with decreasing normalized selfishness, while the cooperative system without the blockchain-empowered credit system, regardless of whether or not our heuristic algorithm is used, builds up more and more selfishness. Although the performance of the heuristic-credit case on the average task execution time is around 20% worse than that without our credit system, the level of selfishness of the case without our credit system is more than three-times higher. Therefore, the adoption of our credit system is highly recommended to enforce fairness in the network.

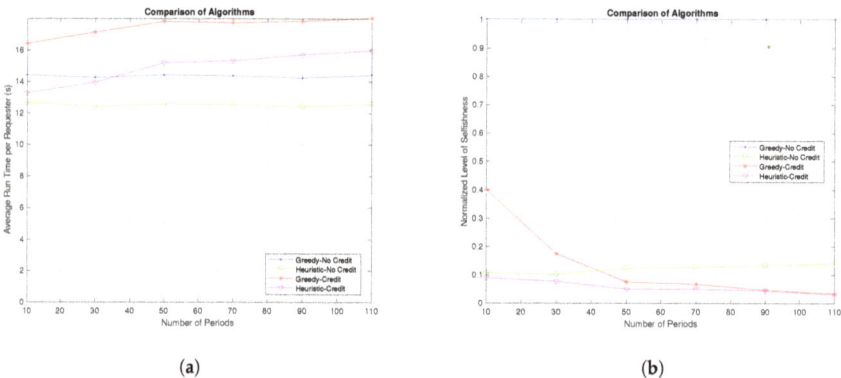

Figure 8. Effect of ongoing number of periods. (**a**) Effect on average task execution time; (**b**) effect on the level of selfishness.

7. Discussion

As shown in Section 6, our blockchain-empowered D2D computational resource sharing system is able to achieve lower average task execution time while enforcing fairness among users. Possible application scenarios include computation virtualization [32] in the D2D network. The authors in [33] presented the feasibility of computational resource virtualization within a personal cloud so that a weak device can utilize computational resources from stronger devices for graphics rendering and other applications. The authors clearly claimed one of the key challenges to be the dynamic nature of a personal cloud caused by the mobility of the user. With our proposed system in this paper, we can extend the idea of computational resource virtualization presented in [33] to a D2D network. A cooperative computational task may be executed in a virtual machine and migrated to different devices in the D2D network. Although virtualization introduces performance and management overhead, the flexibility it can bring to network resource management still makes it very appealing. The fact that virtual machines can be migrated to a different physical host while keeping applications alive makes the computational resource virtualization and sharing in the D2D network even more fascinating [34]. Furthermore, our system enforces fairness among users by incorporating the blockchain-based credit system.

Apart from the pros and cons explained in the Section 6, the benefits of adopting the blockchain-base credit system for our proposed D2D computational resource sharing system are as follows:

- Decentralized and trustless: The blockchain is a public ledger of all transactions in the network. This public ledger is maintained by all participating nodes, and this consensus mechanism makes central authority unnecessary. Therefore, blockchain technology enables a decentralized and trustless network where peers do not need any trusted third party to interact with each other. Note that the supernode in our system mainly works to assign computational resource sharing in an efficient and fair way by considering user mobility and credit balances. Our supernode also provides a public transaction pool as a reference to system users, but the supernode in no way participates in the mining operations. All mining operations are performed by system users in the D2D network.
- Autonomous: Blockchain technology can enable devices in the D2D network to communicate with each other and perform transactions autonomously, since each device can assess the blockchain and a trusted intermediary is not needed. Again, although the supernode helps system users by assigning cooperation tasks in an efficient and fair way, the credit balance system is not controlled by the supernode and remains autonomous.

Our system also faces a few challenges as follows:

- Efficiency: Since all miners in the network perform the same computations trying to get the next block reward from the blockchain, there remains efficiency concerns. In our proposed blockchain-empowered D2D computational resource sharing system, the block reward is credits that could be used to exchange for computational resources, which can also be granted when helping peers computing in the D2D network. Therefore, users in the system are not merely encouraged to compete for the block rewards, but also encouraged to assist other peers, which enhances the efficiency of users' idle computational resources.
- Privacy: Because the blockchain is a public ledger and any node can see all transactions in the network, privacy concerns remain for the transacting parties.
- Interference in the D2D network: In this work, we de-emphasized the effect of mutual interference in the D2D network due to the limited D2D communications' duration and the coordination from the supernode. To realize a more realistic model, we will elaborate on how mutual interference can be tackled by supernode coordination in future works.

8. Conclusions

In this work, we build a blockchain-empowered credit system on top of the connectivity-aware computational resource sharing system in the D2D network. A supernode at the base station, with knowledge of user mobilities and thus the probability model of device connectivities, will perform task scheduling to reduce average task execution time for requesters in the network and enhance user quality of experience. Based on the blockchain-based credit system, selfish users who only want to get help from peers, but not contribute, will not be assigned any helper assistance if their balance is not sufficient. The supernode also possesses a publicly accessible transaction pool for miners' reference on building up a trustworthy blockchain network. Simulation results based on a realistically examined mobility model show that our system substantially reduces average task execution time for requesters in the D2D network. Sacrificing a minor amount of average task execution time allows the system to remain at a rather low level of selfishness. To enforce fairness and encourage users, the adoption of our credit system is highly recommended. With the help of blockchain technology, our system becomes more favourable for users by providing incentives to helpers and enforces fairness among users.

Acknowledgments: This work is supported by funding from the Natural Sciences and Engineering Research Council of Canada.

Author Contributions: Zhen Hong conceived of, designed and performed the experiments. Zhen Hong analysed the data. Zehua Wang and Wei Cai contributed reagents/materials/analysis tools. Zhen Hong, Zehua Wang and Wei Cai wrote the paper. Victor C. M. Leung supervised the work.

Conflicts of Interest: The authors declare no conflict of interest.

Abbreviations

The following abbreviations are used in this manuscript:

D2D	device-to-device
QoE	quality of experience
IoT	Internet of Things
BS	base station
LoS	level of selfishness
NP-hard	non-deterministic polynomial-time hard
CPU	central processing unit
4G	Fourth Generation
LTE	Long Term Evolution

References

1. Tai, Y.M.; Ku, Y.C. Will stock Investors use mobile stock trading? A benefit-risk assessment based on a modified utaut model. *J. Electron. Commer. Res.* **2013**, *14*, 67–84.
2. Cai, W.; Leung, V.C.M.; Chen, M. Next Generation Mobile Cloud Gaming. In Proceedings of the 2013 IEEE Seventh International Symposium on Service-Oriented System Engineering, Redwood City, CA, USA, 25–28 March 2013; pp. 551–560.
3. PlayStation Now. Available online: https://www.playstation.com (accessed on 30 September 2017).
4. GameFly. Available online: https://www.gamefly.com (accessed on 30 September 2017).
5. Yi, S.; Li, C.; Li, Q. A Survey of Fog Computing: Concepts, Applications and Issues. In Proceedings of the 2015 Workshop on Mobile Big Data, Hangzhou, China, 21 June 2015; ACM: New York, NY, USA, 2015; pp. 37–42.
6. Dave Chaffey. Mobile Marketing Statistics Compilation. Available online: http://www.smartinsights.com/mobile-marketing/mobile-marketing-analytics/mobile-marketing-statistics/?new=1 (accessed on 30 September 2017).
7. Shi, C.; Lakafosis, V.; Ammar, M.H.; Zegura, E.W. Serendipity: Enabling Remote Computing Among Intermittently Connected Mobile Devices. In Proceedings of the Thirteenth ACM International Symposium on Mobile Ad Hoc Networking and Computing, Hilton Head, SC, USA, 11–14 June 2012; ACM: New York, NY, USA, 2012; pp. 145–154.
8. Fahim, A.; Mtibaa, A.; Harras, K.A. Making the Case for Computational Offloading in Mobile Device Clouds. In Proceedings of the 19th Annual International Conference on Mobile Computing & Networking, Miami, FL, USA, 30 September–4 October 2013; ACM: New York, NY, USA, 2013; pp. 203–205.

9. Shi, C.; Ammar, M.H.; Zegura, E.W.; Naik, M. Computing in Cirrus Clouds: The Challenge of Intermittent Connectivity. In Proceedings of the First Edition of the MCC Workshop on Mobile Cloud Computing, Helsinki, Finland, 17 August 2012; ACM: New York, NY, USA, 2012; pp. 23–28.

10. Hong, Z.; Wang, Z.; Cai, W.; Leung, V.C.M. Connectivity-Aware Task Outsourcing and Scheduling in D2D Networks. In Proceedings of the 2017 26th International Conference on Computer Communication and Networks (ICCCN), Vancouver, BC, Canada, 31 July–3 August 2017; pp. 1–9.

11. Hong, Z.; Cai, W.; Wang, X.; Leung, V.C.M. Reputation-based multiplayer fairness for ad-hoc cloudlet-assisted cloud gaming system. In Proceedings of the 2014 International Conference on Smart Computing, Hong Kong, China, 3–5 November 2014; pp. 89–96.

12. Pilkington, M. Blockchain Technology: Principles and Applications. In *Research Handbook on Digital Transformations*, Xavier Olleros, F., Zhegu, M., Eds.; Edward Elgar: Northampton, MA, USA, 2016.

13. Ra, M.R.; Paek, J.; Sharma, A.B.; Govindan, R.; Krieger, M.H.; Neely, M.J. Energy-delay Tradeoffs in Smartphone Applications. In Proceedings of the 8th International Conference on Mobile Systems, Applications, and Services, San Francisco, CA, USA, 15–18 June 2010; ACM: New York, NY, USA, 2010; pp. 255–270.

14. Ding, A.Y.; Han, B.; Xiao, Y.; Hui, P.; Srinivasan, A.; Kojo, M.; Tarkoma, S. Enabling energy-aware collaborative mobile data offloading for smartphones. In Proceedings of the 2013 IEEE International Conference on Sensing, Communications and Networking (SECON), New Orleans, LA, USA, 24–27 June 2013; pp. 487–495.

15. Mehmeti, F.; Spyropoulos, T. Is it worth to be patient? Analysis and optimization of delayed mobile data offloading. In Proceedings of the IEEE INFOCOM 2014—IEEE Conference on Computer Communications, Toronto, ON, Canada, 27 April–2 May 2014; pp. 2364–2372.

16. Asadi, A.; Mancuso, V. WiFi Direct and LTE D2D in action. In Proceedings of the 2013 IFIP Wireless Days (WD), Valencia, Spain, 13–15 November 2013; pp. 1–8.

17. Chen, C.A.; Won, M.; Stoleru, R.; Xie, G.G. Energy-Efficient Fault-Tolerant Data Storage and Processing in Mobile Cloud. *IEEE Trans. Cloud Comput.* **2015**, *3*, 28–41.

18. Boccardi, F.; Heath, R.W.; Lozano, A.; Marzetta, T.L.; Popovski, P. Five disruptive technology directions for 5G. *IEEE Commun. Mag.* **2014**, *52*, 74–80.

19. Gao, C.; Gutierrez, A.; Rajan, M.; Dreslinski, R.G.; Mudge, T.; Wu, C.J. A study of mobile device utilization. In Proceedings of the 2015 IEEE International Symposium on Performance Analysis of Systems and Software (ISPASS), Philadelphia, PA, USA, 29–31 March 2015; pp. 225–234.

20. Wang, Z.; Shah-Mansouri, H.; Wong, V. How to Download More Data from Neighbors? A Metric for D2D Data Offloading Opportunity. *IEEE Trans. Mob. Comput.* **2016**, *16*, 1658–1675.

21. Ye, F.; Zheng, Z.; Chen, C.; Zhou, Y. DC-RSF: A Dynamic and Customized Reputation System Framework for Joint Cloud Computing. In Proceedings of the 2017 IEEE 37th International Conference on Distributed Computing Systems Workshops (ICDCSW), Atlanta, GA, USA, 5–8 June 2017; pp. 275–279.

22. Merkle, R.C. A Digital Signature Based on a Conventional Encryption Function. In *Advances in Cryptology—CRYPTO '87: Proceedings*; Pomerance, C., Ed.; Springer: Berlin, Germany, 1988; pp. 369–378.

23. Jakobsson, M.; Juels, A. Proofs of Work and Bread Pudding Protocols(Extended Abstract). In *Secure Information Networks: Communications and Multimedia Security IFIP TC6/TC11 Joint Working Conference on Communications and Multimedia Security (CMS'99), Leuven, Belgium, 20–21 September 1999*; Preneel, B., Ed.; Springer: Boston, MA, USA, 1999; pp. 258–272.

24. Norris, J.R. *Markov Chains*; Number 2; Cambridge University Press: Cambridge, UK, 1998.

25. Gerber, R.; Hong, S. Slicing Real-time Programs for Enhanced Schedulability. *ACM Trans. Program. Lang. Syst.* **1997**, *19*, 525–555.

26. Dean, J.; Ghemawat, S. MapReduce: Simplified Data Processing on Large Clusters. *Commun. ACM* **2008**, *51*, 107–113.

27. Kellerer, H.; Pferschy, U.; Pisinger, D. Introduction to NP-Completeness of Knapsack Problems. In *Knapsack Problems*; Springer: Berlin, Germany, 2004; pp. 483–493.

28. MATLAB. Available online: https://www.mathworks.com/help/optim/ug/linprog.html (accessed on 30 September 2017).

29. Yang, Y. An Efficient Polynomial Interior-Point Algorithm for Linear Programming. 2013. Available online: https://arxiv.org/abs/1304.3677 (accessed on 30 September 2017).

30. Scott, J.; Gass, R.; Crowcroft, J.; Hui, P.; Diot, C.; Chaintreau, A. CRAWDAD Dataset Cambridge/haggle (v. 2006-09-15). 2006. Available online: https://crawdad.org/cambridge/haggle/20060915/imote/ (accessed on 30 September 2017).

31. Chang, P.C.; Liu, C.H.; Lin, J.L.; Fan, C.Y.; Ng, C.S. A neural network with a case based dynamic window for stock trading prediction. *Exp. Syst. Appl.* **2009**, *36*, 6889–6898.

32. Kumar, K.; Liu, J.; Lu, Y.H.; Bhargava, B. A Survey of Computation Offloading for Mobile Systems. *Mob. Netw. Appl.* **2013**, *18*, doi:10.1007/s11036-012-0368-0.

33. Wu, X.; Wang, W.; Lin, B.; Miao, K. Composable IO: A novel resource sharing platform in personal Clouds. *J. Supercomput.* **2012**, *61*, doi:10.1007/s11227-011-0663-8.

34. Clark, C.; Fraser, K.; Hand, S.; Hansen, J.G.; Jul, E.; Limpach, C.; Pratt, I.; Warfield, A. Live migration of virtual machines. In Proceedings of the NSDI'05 2nd conference on Symposium on Networked Systems Design & Implementation, Berkeley, CA, USA, 2–4 May 2005; Volume 2, pp. 273–286.

future internet

MDPI

Article

Energy-Efficient Resource and Power Allocation for Underlay Multicast Device-to-Device Transmission

Fan Jiang [1,*], Honglin Wang [1], Hao Ren [1] and Shuai Xu [2]

[1] Shaanxi Key Laboratory of Information Communication Network and Security, Xi'an University of Posts and Telecommunications, Xi'an 710121, China; wanghonglin928@gmail.com (H.W.); 18591970531@163.com (H.R.)

[2] College of Liberal Arts; Xi'an University of Finance and Economics, Xi'an 710121, China; xushuai.8@163.com

[*] Correspondence: jiangfan@xupt.edu.cn

Received: 20 October 2017; Accepted: 8 November 2017; Published: 14 November 2017

Abstract: In this paper, we present an energy-efficient resource allocation and power control scheme for D2D (Device-to-Device) multicasting transmission. The objective is to maximize the overall energy-efficiency of D2D multicast clusters through effective resource allocation and power control schemes, while considering the quality of service (QoS) requirements of both cellular users (CUs) and D2D clusters. We first build the optimization model and a heuristic resource and power allocation algorithm is then proposed to solve the energy-efficiency problem with less computational complexity. Numerical results indicate that the proposed algorithm outperforms existing schemes in terms of throughput per energy consumption.

Keywords: Device-to-Device (D2D) communication; multicast transmission; energy-efficiency

---

1. Introduction

With the emergence and popularity of the Internet of Things (IoT) [1,2], billions of devices will be connected and serviced by current wireless networks. In particular, the local area service of popular content sharing is one of the main reasons for this tremendous growth. Such unprecedented growth of data has brought great pressure to current network architectures and technologies. Under this circumstance, the direct connectivity between mobile devices, namely, Device-to-Device (D2D) communication underlay cellular networks, emerges as a potential component for the fifth generation (5G) mobile networks [3].

The concept of underlay D2D multicast transmission refers to the high spectrum-efficient D2D multicast transmission scenario (from the cluster head (CH) to multiple member user equipment (UEs)) which reuses the resource of existing cellular links [4]. By exploiting the inherent broadcast nature of wireless channels, D2D multicast transmission provides an effective solution to offload the heavy data traffic to D2D links, which not only mitigates the burden of the base station (BS) but also increases the spectrum efficiency of the network [5]. Concerning with current researches about D2D multicast transmission, most works mainly focus on utilizing D2D multicast to improve the system spectrum efficiency, or to provide offloading function while the energy efficiency of D2D multicast transmission has often been omitted [6–11]. In this paper, based on the existing contributions, we propose an energy-efficient resource allocation and power control strategy for D2D multicast transmission scenario. Specifically, in order to maximize the overall energy-efficiency of D2D multicast clusters, we first formulate the energy-efficiency optimization problem which is a non-convex problem. Then, we propose a heuristic resource allocation and power control algorithm, which brings computational complexity compared with the conventional exhaustive searching based algorithms. The proposed scheme has better performance with respect to energy-efficiency.

The rest of the paper is organized as follows. In Section 3, we formulate the network model and illustrate the energy-efficiency maximization problem. Section 4 investigates the resource and power allocation problem with energy-efficient consideration and a heuristic algorithm is then proposed. Simulation results and analysis are given in Section 5. Finally, Section 6 concludes the paper.

2. Related Work

Recently, investigations about underlay D2D multicast transmission mainly concentrate on how to mitigate the reuse interference introduced by D2D transmission [6–8]. For example, in order to maximize the total throughput of CUs and D2D clusters in a cellular cell, Meshgi et al. [6] studied a joint channel and power allocation strategy. The work utilizes the maximum weight bipartite matching method to find the optimal resource allocation and power allocation between CUs and D2D pairs. Bhardwaj et al. [7] proposed a scheme to minimize the interference among D2D links and CUs through a resource allocation scheme. The object is to maximize the total throughput of CUs and D2D users through a joint power and channel allocation scheme. Kitagawa et al. [8] proposed an efficient transmitter user selection algorithm, which improves system capacity while minimizing the impact of interference among D2D multicast communications. However, it can be easily seen that most of the above works focus on how to improve the system spectrum efficiency of D2D multicasting while the energy-efficiency aspect of D2D multicasting is not properly addressed.

On the other hand, although some studies already deal with the energy-efficiency aspect of D2D transmission, their research focuses are different from ours. In [9], D2D multicast transmission is suggested to perform the computation offloading task for interactive applications. The proposed algorithm aims at minimizing the energy consumption of each mobile terminal other than the overall energy-efficiency. In [10], a joint power and resource allocations scheme is proposed for D2D underlay multicast communication. This work mainly focuses on how to accommodate more D2D multicast groups while minimizing the total terminal transmission power. In [11], a D2D crowd framework for 5G mobile edge computing is proposed. The authors first introduced the concept of D2D crowd framework, then propose a graph matching-based optimal task assignment policy to address the energy efficient D2D task assignment problem. By taking the energy constraint into account, this work mainly deals with D2D crowd task assignment problem, while the energy-efficiency of the D2D clusters is not considered.

3. System Model and Problem Formulation

Without loss of generality, we assume that several D2D clusters have already been formed either by the BS coordinately or by the cluster head in a distributed way, as shown in Figure 1. In fact, D2D cluster formation methods have been widely discussed in existing works, such as [12]. Hence, this paper mainly focuses on cluster-based D2D multicast transmission underlying uplink cellular networks. As illustrated in Figure 1, supposes there are M D2D multicast clusters who share uplink resource blocks (RBs) with N CUs. We use $m, m \in \mathcal{M} = \{1, 2, \ldots, M\}$ to index m-th D2D cluster, where $\mathcal{M}$ is the set of D2D multicast clusters. To clarify computation, the resource allocation for CUs is assumed to be pre-determined (e.g., the n-th RB is allocated to the n-th CU). Let $n, n \in \mathcal{N} = \{1, 2, \ldots, N\}$ indicates the n-th CU and also the RB it occupies, where $\mathcal{N}$ represents the set of CUs. This paper mainly focuses on the matching of D2D multicast clusters and CU as well as the power control policies correspondingly. Within each D2D multicast cluster, the cluster head serves as the D2D transmitter and the cluster members are D2D receivers. Suppose $\mathcal{K}_m$ is used to represent the set of receivers in the m-th D2D multicast cluster, where $|\mathcal{K}_m|$ denotes the total number of receivers in the m-th D2D cluster. When $|\mathcal{K}_m| = 1$, the transmission scenario becomes unicast transmission.

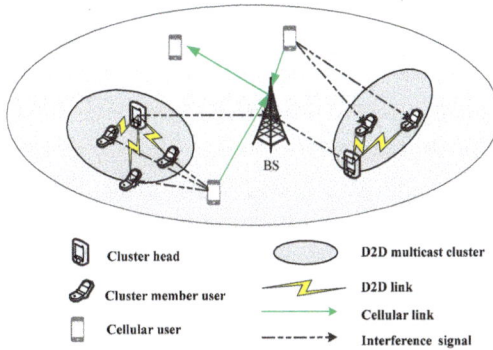

Figure 1. The considered device-to-device (D2D) multicast transmission network model. BS: base station.

We define a set of binary variables $x_{m,n}$. If the m-th D2D cluster reuses the RB of n-th CU, then $x_{m,n} = 1$, otherwise, $x_{m,n} = 0$. Assume each D2D multicast cluster is allowed to reuse at most one RB. Then we have

$$\begin{cases} \sum_{n=1}^{N} x_{m,n} \leq 1, \quad \forall m \in \mathcal{M} \\ \sum_{m=1}^{M} x_{m,n} \leq 1, \quad \forall n \in \mathcal{N} \end{cases} \tag{1}$$

Equation (1) indicates two aspects. One is that each D2D multicast cluster is allowed to reuse at most one RB of a CU; the other is the RB of a CU can only be reused by at most one D2D cluster. Consequently, assume the n-th RB is reused by m-th D2D cluster, the channel quality of the k-th member UE who act as a receiver in m-th D2D cluster is given by

$$\gamma_k^{D2D_m} = \frac{G_{m,k}^{D2D}}{\sum_{n=1}^{N} x_{m,n} P_n^{CU} G_{n,k}^{C2D} + \sigma_n^2}, \forall m \in \mathcal{M}, k \in \mathcal{K}_m, n \in \mathcal{N} \tag{2}$$

where $\gamma_k^{D2D_m}$ denotes the channel quality of k-th cluster member in m-th D2D multicast cluster. P_n^{CU} represents the transmit power of CU n who shares uplink RB n together. $G_{m,k}^{D2D}$ and $G_{n,k}^{C2D}$ stand for the channel gain between the cluster head and cluster member k in the m-th D2D cluster, between interfering CU n to member UE k, respectively. σ_n^2 denotes the noise power.

Similarly, the channel quality of a CU n is calculated as:

$$\gamma_n^{CU} = \frac{G_{n,BS}^{CU}}{\sum_{m=1}^{M} x_{m,n} P_m^{D2D_m} G_{m,BS}^{D2C} + \sigma_n^2}, \forall m \in \mathcal{M}, n \in \mathcal{N} \tag{3}$$

Here, $P_m^{D2D_m}$ stands for the transmit power of the cluster head in m-th D2D cluster who reuses the RB of CU n. $G_{n,BS}^{CU}$ and $G_{m,BS}^{D2C}$ represent the link gain from CU n to the BS and from co-channel cluster head to the BS, respectively.

According to [6], it is commonly assumed that the transmission rate of a multicast is determined by the user with the worst channel condition. Combined with the scenario shown in Figure 1, for the m-th D2D cluster, the transmission rate achieved at RB n is given by

$$\gamma_{m,n}^{D2D} = \min_{\forall k \in \mathcal{K}_m} \gamma_k^{D2D_m} \tag{4}$$

Therefore, we can calculate the normalized transmission rate of the m-th D2D cluster as

$$r_m^{D2D} = \sum_{n=1}^{N} x_{m,n} \log_2 \left(1 + P_m^{D2D_m} \gamma_{m,n}^{D2D} \right) \tag{5}$$

As a result, the transmission rate of m-th D2D cluster is expressed as

$$R_m^{D2D} = |\mathcal{K}_m| r_m^{D2D} \tag{6}$$

Similarly, we can formulate the normalized transmission rate for a CU n which use RB n as

$$R_n^{CU} = \log_2 \left(1 + P_n^{CU} \gamma_n^{CU} \right) \tag{7}$$

In order to guarantee the quality of service (QoS) requirement of each UE, a threshold is set. Specifically, for the m-th D2D multicast cluster, the above requirement is expressed as:

$$P_m^{D2D_m} \gamma_m^{D2D} \geq \Gamma_{min}^{D2D} \tag{8}$$

The above expression is explained as in order to ensure reliable transmission, the signal-to-interference-and-noise ratio (SINR) of the user who has the worst channel condition among a D2D multicast cluster should be above certain threshold. Again, for a CU, this requirement is set as

$$P_n^{CU} \gamma_n^{CU} \geq \Gamma_{min}^{CU} \tag{9}$$

where Γ_{min}^{D2D} and Γ_{min}^{CU} represent the minimum SINR threshold to ensure reliable D2D multicast and cellular transmission specified by the system.

Combined with the maximum transmit power constraints for CUs and D2D clusters and by substituting expressions (2) and (3) into (8) and (9), the transmit power range of a cluster head and a CU is expressed as

$$
\begin{cases}
\min \dfrac{\dfrac{\Gamma_{min}^{D2D}}{G_{m,k}^{D2D}}}{\sum\limits_{n=1}^{N} x_{m,n} P_n^{CU} G_{n,k}^{C2D} + \sigma_n^2} \leq P_m^{D2D_m} \leq P_{max}^{D2D}, \forall m \in \mathcal{M}, k \in \mathcal{K}_m \\[4ex]
\dfrac{\dfrac{\Gamma_{min}^{CU}}{G_{n,BS}^{CU}}}{\sum\limits_{m=1}^{M} x_{m,n} P_m^{D2D_m} G_{m,BS}^{D2C} + \sigma_n^2} \leq P_n^{CU} \leq P_{max}^{CU}, \forall m \in \mathcal{M}, n \in \mathcal{N}
\end{cases}
\tag{10}
$$

where P_{max}^{CU} and P_{max}^{D2D} represent the maximum allowed transmit power of a CU and a D2D user, respectively. From expression (10), we can deduce that besides the channel condition factor, the transmit power of the cluster head as well as the CU who shares the RB resource with a D2D multicast cluster are intertwined. In order to guarantee reliable transmission, the transmit power of different kinds of UEs should be considered.

Given the limited energy capacity of each device, the objective of this paper is to maximize the overall energy efficiency of all D2D multicast clusters, which is also a hottest research aspect. According to [13], the energy-efficiency (EE) of a single D2D multicast cluster m can be expressed as

$$\eta_m = \frac{R_m^{D2D}}{P_m^{D2D_m}} \tag{11}$$

Combing Equations (1)–(11) and since there are M D2D multicast clusters, the energy efficiency of all the D2D multicast clusters is defined as the ratio of total D2D multicast transmission data rates to the overall consumed power of all clusters. Consequently, the energy efficiency optimization problem can be expressed as

$$\max_{\forall m \in \mathcal{M}} \eta_{all} = \frac{\sum\limits_{m=1}^{M} R_m^{D2D}}{\sum\limits_{m=1}^{M} P_m^{D2D_m}} \tag{12}$$

s.t.

$$P_m^{D2D_m} \gamma_m^{D2D} \geq \Gamma_{min}^{D2D}, \forall m \in \mathcal{M}, n \in \mathcal{N} \tag{12a}$$

$$P_n^{CU} \gamma_n^{CU} \geq \Gamma_{min}^{CU} \tag{12b}$$

$$\sum_{n=1}^{N} x_{m,n} \leq 1, \forall m \in \mathcal{M} \tag{12c}$$

$$\sum_{m=1}^{M} x_{m,n} \leq 1, \forall n \in \mathcal{N} \tag{12d}$$

$$\frac{\Gamma_{min}^{D2D}}{\frac{G_{m,k}^{D2D}}{\min \sum\limits_{n=1}^{N} x_{m,n} P_n^{CU} G_{n,k}^{C2D} + \sigma_n^2}} \leq P_m^{D2D_m} \leq P_{max}^{D2D}, \forall m \in \mathcal{M} \tag{12e}$$

$$\frac{\Gamma_{min}^{CU}}{\frac{G_{n,BS}^{CU}}{\sum\limits_{m=1}^{M} x_{m,n} P_m^{D2D_m} G_{m,BS}^{D2C} + \sigma_n^2}} \leq P_n^{CU} \leq P_{max}^{CU}, \forall m \in \mathcal{M}, n \in \mathcal{N} \tag{12f}$$

where η_{all} represents the overall energy-efficiency of D2D clusters. Constraints (12a) and (12b) define the minimum SINR requirement. Constraints (12c) and (12d) ensure that each D2D cluster reuses the RBs of CU at most once. Constraints (12e) and (12f) ensure that the transmit power both D2D users and CUs fall into certain range.

According to Equation (12), in order to improve the total energy-efficiency, the possible solution is either to decrease the transmit power or to increase the aggregate transmission rate. This is interpreted as we have to find the optimal resource reuse relationship $x_{m,n}$ between D2D clusters and CUs, as well as to determine the optimal transmit power of both CUs and D2D clusters which also guarantees the minimum SINR requirement.

As a matter of fact, when D2D multicast transmission undelaying with a cellular network, the resource allocation method and power allocation strategy are actually interacted with each other [14]. Once the resource reuse relationship between CUs and D2D clusters varies, the transmit power of each UE will also be influenced owing to changing interference condition. On the other hand, if the transmit power of different UE alters, the interference condition between co-channel CUs and D2D clusters will also change, which conversely affects the resource assignment results. In addition, the existence of integer assignment variable $x_{m,n}$ makes the optimization problem more complicated.

Consequently, the above optimization problem in (12) is a non-convex optimization problem which is proved to be a NP-Hard problem and there are no efficient solutions [15]. Moreover, when the problem size increases, the computational complexity also increases exponentially. A possible solution for above problem might be using the bipartite matching based optimal resource allocation scheme, as suggested by [16]. However, such scheme is actually based on exhaustive searching method, which results in high computational complexity. In order to deal with this challenge, we propose a heuristic resource allocation and power control algorithm to balance the system performance and complexity.

4. Proposed Heuristic Energy-Efficient Resource and Power Allocation Algorithm

According to (12), we can deduce that in order to improve energy-efficiency of all the D2D multicast clusters, we have to either decrease the transmit power or to increase the aggregate transmission rate. However, due to the fraction form of the objective function built in (12), it is a non-convex optimization problem. Hence, it is difficult to obtain the optimal solution directly. In the following part, we will first solve the overall energy-efficiency optimization problem by adopting a heuristic resource allocation strategy.

Firstly, we consider how to increase the numerator of expression (12). From the resource assignment point of view, in order to improve the aggregate D2D throughput, higher values of SINR are desirable. From constraints (12a)–(12f), we can infer that a smaller value of $G_{n,m}^{C2D}$ means less interference from co-channel CU n to D2D cluster m, which will result in higher γ_m^{D2D} and D2D throughput. Hence, the fundamental idea of the proposed scheme is to pick up the CU who generates less interference to the co-channel D2D cluster. To achieve that target, we build a channel state information (CSI) matrix $G_{M \times N}^{C2D}$, which is composed of estimated channel gain information of each D2D cluster from the interfering CU respectively, where $G_{M \times N}^{C2D}$ is expressed as

$$
G_{M \times N}^{C2D} =
\begin{bmatrix}
G_{1,1}^{C2D}, G_{2,1}^{C2D}, ..., G_{N,1}^{C2D} \\
G_{1,2}^{C2D}, G_{2,2}^{C2D}, ..., G_{N,2}^{C2D} \\
\vdots \\
G_{1,M}^{C2D}, G_{2,M}^{C2D}, ..., G_{N,M}^{C2D}
\end{bmatrix}
\tag{13}
$$

In the matrix $G_{M \times N}^{C2D}$, each element stands for the CSI value from the co-channel CU to the corresponding D2D multicast cluster. We assume that the CSI between each CU and each D2D cluster can be obtained by each cluster head individually. Actually, such information can be initially obtained through information change between each cluster head and D2D receivers. Then, the CSI information between each cluster and each CU can be gathered at the BS side through control information exchange. Consequently, we can find the minimum $G_{n,m}^{C2D}$, $n \in \mathcal{N}, m \in \mathcal{M}$ in each row so as to pair up the CU which brings the least interference to the D2D multicast cluster. By doing the same procedure for each D2D multicast cluster, the optimal resource assignment between CU and the corresponding D2D multicast cluster can be decided.

After picking out the cellular resource for each D2D cluster, the next step is to decrease the denominator of expression (12). This is interpreted as to determine the minimum transmit power of each CU and D2D cluster respectively, which also satisfies the minimum SINR threshold requirement. Suppose a CU n shares RB resource with a D2D cluster m after resource pairing process. By using the Equations (8) and (9) and substituting γ_n^{CU} and γ_m^{D2D} from expression (3) and (4) respectively, we can determine the optimal transmit power of a D2D cluster and a CU which can guarantee reliable transmission as follows

$$
\begin{cases}
P_m^{D2D_m} = \max\limits_{\forall m \in \mathcal{M}, k \in \mathcal{K}_m, n \in \mathcal{N}} \dfrac{\Gamma_{min}^{D2D} \Gamma_{min}^{CU} G_{n,k}^{C2D} \sigma_n^2 + \Gamma_{min}^{D2D} G_{n,BS}^{CU} \sigma_n^2}{G_{m,k}^{D2D} G_{n,BS}^{CUE} - \Gamma_{min}^{D2D} \Gamma_{min}^{CU} G_{n,k}^{C2D} G_{m,BS}^{D2C}} \\[6mm]
P_n^{CU} = \max\limits_{\forall m \in \mathcal{M}, k \in \mathcal{K}_m, n \in \mathcal{N}} \dfrac{\Gamma_{min}^{D2D} \Gamma_{min}^{CU} G_{m,BS}^{D2C} \sigma_n^2 + \Gamma_{min}^{CU} G_{m,k}^{D2D} \sigma_n^2}{G_{m,k}^{D2D} G_{n,BS}^{CU} - \Gamma_{min}^{D2D} \Gamma_{min}^{CU} G_{n,k}^{C2D} G_{m,BS}^{D2C}}
\end{cases}
\tag{14}
$$

Combined with the expressions given in (14), if the calculated transmit power of both CUs and D2D users obtained in (14) do not exceed the maximum allowed transmit power, CU n is finally chosen as the resource sharing partner for D2D cluster m. By substituting (14) into the optimization function (12), we can determine the minimum transmit power of different D2D multicast cluster so as to maximize the overall energy-efficiency of D2D clusters. On the contrary, if the calculated transmit power of both CUs and D2D users exceed the maximum allowed values, then CU m will be removed from the available resource assignment list and we will try the next available CU according to the matrix $G_{M \times N}^{C2D}$. The pseudo code of the proposed energy-aware resource allocation and power control algorithm is given in Algorithm 1.

Algorithm 1 A heuristic energy-efficient resource and power allocation scheme for D2D multicast transmission

---

1:   $\mathcal{M}$: List of D2D clusters
2:   $\mathcal{N}$: List of CUs
3:   Construct the matrix $\mathbf{G}^{C2D}_{M \times N}$ according to (13),
4:   $i = 1, j = 1,$
5:   **while** $\mathcal{M} \neq \varnothing, i \leq N$ and $j \leq M$ **do**
6:   $G^{C2D}_{n,m} = \text{argmin} G^{C2D}_{i,j}, i \in \mathcal{N}, j \in \mathcal{M},$
7   Record the value of n and m,
8:   Find $P^{D2D_m}_m$ and P^{CU}_n from (14),
9:      **if** $P^{D2D_m}_m \leq P^{D2D}_{\max}$ and $P^{CU}_n \leq P^{CU}_{\max}$ **then**
10:       D2D cluster m shares resource with CU n,
11:      Substitute $P^{D2D_m}_m$ and P^{CU}_n into the maximization problem built in (12), find the minimum transmit
power;
12       $j = j + 1,$
13:   **else**
14:       $i = i + 1,$
15:   **end if**
16: **end while**
17: Compute η_{all} according to (12)

In the proposed scheme, in order to maximize the overall energy-efficiency of D2D multicast clusters, we first search for the optimal RB assignment between CUs and D2D clusters. This leads to increased total data rates of D2D multicast transmission. Then, combined with the minimum SINR requirement, we find out the minimum transmit power of both CUs and D2D clusters to ensure reliable D2D multicast transmission. Hence, the maximization problem built in expression (12) is solved by a two-step way, where the computational complexity of the proposed strategy is $(MN) + (M \times f|\mathcal{K}_m|)$. This is because in the worst case, the maximization problem will be solved in M times. Here, $f|\mathcal{K}_m|$ represents the size of each D2D multicast cluster. It can be seen that, compared with exhaustive searching based methods, (e.g., such as [16], where the complexity is $(M^3) + (M \times N \times f|\mathcal{K}_m|)$), the proposed heuristic solution can considerably reduce the computational complexity.

5. Simulation Results

In this section, numerical results are provided to demonstrate the performance of our proposed strategy. We use the clustered distribution model adopted in [17], where a 400 m × 400 m square area is used to simulate the network. Cluster heads are randomly distributed in the simulation area according to the uniform distribution and the D2D users are randomly distributed in the corresponding multicast cluster. The distance-based path loss and shadowing fading are considered for the transmission channel. We still consider the scenario that D2D clusters and CUs share uplink cellular RBs together. Suppose that all available resource is divided into RBs and each CU is allocated with one RB at each scheduling slot. Other related simulation parameters are listed shown in Table 1.

In order to demonstrate the performance of the proposed energy-aware resource and power allocation scheme, three other different algorithms are considered. The first one is the QoS-aware resource allocation scheme proposed in [7], which aims at minimizing the interference among D2D multicast cluster and CU through resource allocation. Moreover, the power control policy in [7] assumes that both CU and D2D cluster transmit at the maximum power when the channel condition is good enough. The second one is the cluster based scheme proposed in [18], which employs social information to facilitate file transfer process. In the absence of resource pairing scheme between D2D cluster and CU, the resource assignment between D2D cluster and CU in [18] is assumed to be chosen randomly while power control method is not applied. The third one is the energy-efficient scheme proposed in [14], which aims at improving the energy efficiency multicast transmission through proper power control. The power control principle of [14] is similar to our proposed scheme which tries to allocate more transmit power to D2D pairs when the channel condition becomes good. For clarity, our proposed scheme is referred to as the "Proposed energy-aware" scheme.

Table 1. The simulation parameters. UE: user equipment; RB: resource blocks; SINR: signal-to-interference-and-noise ratio; CU: cellular users; D2D: Device-to-Device.

Parameter	Value
Cell Radius	400 m
Total UE number	500, 1000
Spectrum bandwidth	10 MHz
Bandwidth of each RB	180 KHz
The path loss Component (α)	4
Shadowing	Log-normal fading with standard deviation of 8 dB
Noise power spectrum density (σ^2)	−174 dBm/Hz
Minimum SINR Threshold ($\Gamma_{min}^{D2D}, \Gamma_{min}^{CU}$)	10 dB
Maximum transmit power of UE ($P_{max}^{CU}, P_{max}^{D2D}$)	CU: 23 dBm, D2D user: 20 dBm
D2D cluster radius (r)	30~90 m
UE Transmission range	90 m

Figure 2 illustrates the sum energy efficiency of the D2D multicast cluster with the variation of the D2D cluster radius. From Figure 2 we can see that with the increase of the D2D cluster radius, the energy efficiency performance of our proposed scheme is better than other referenced schemes. This is because the channel gain of the D2D link will decrease with the increase of D2D cluster radius. Hence, a larger transmit power is required for the D2D clusters so as to satisfy the SINR threshold constraint. Accompanied with the increase of D2D cluster radius, our proposed scheme gradually increases the transmit power. This results in the decreased energy efficiency of the D2D cluster. Compared with the scheme in [14], our proposed resource allocation method minimizes the interference from CUs to co-channel D2D receivers, which improves the channel gain of D2D links. Furthermore, it is shown that when there are more UEs in the considered scenarios (*N* varies), the overall energy efficiency also increases due to the increased total transmission rate of all D2D multicast clusters.

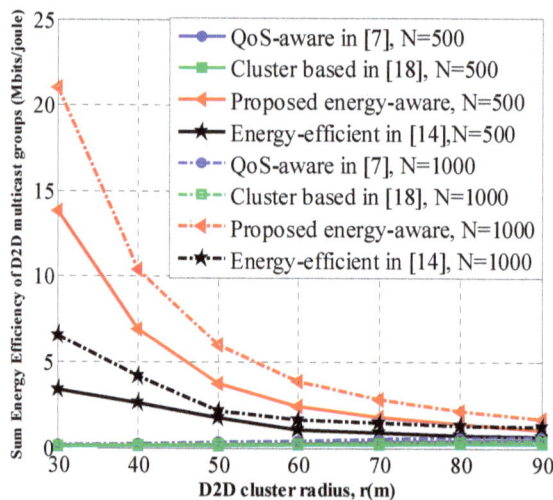

Figure 2. Cluster Energy Efficiency versus D2D cluster radius. QoS: quality of service.

Figure 3 compares the total throughput of D2D multicast clusters of the proposed scheme with three other schemes. From the figure, we can see that the total throughput performance of the proposed scheme is initially inferior to that of [7] and gradually outperforms other schemes with the increase of D2D cluster radius. The reason is twofold. Firstly, our proposed scheme aims at maximizing the EE performance of the D2D clusters, which adopts a smaller transmit power when the radius of the D2D multicast group is small. This explains why the total throughput of our proposed scheme is smaller than that of the scheme in [7], which always adopts the maximum transmit power. However, with the increase of D2D cluster radius, our proposed scheme will gradually increase the transmit power of D2D users while decrease the transmit power of co-channel CUs in order to ensure reliable D2D multicast transmission, which leads to increased total throughput. Secondly, compared with schemes in [14,18], the resource relationship of our proposed scheme is based on minimizing the interference of D2D links. As a result, the CU who brings the least interference will be paired up with the D2D multicast cluster accordingly. This explains why our proposed scheme has better performance. Similarly, it is shown that the sum throughput of all four algorithms increases with the increase of user numbers.

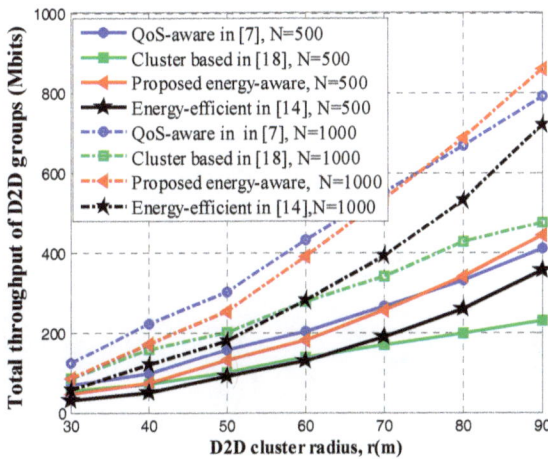

Figure 3. Total Throughput of D2D multicast clusters versus D2D cluster radius.

The impact of energy efficiency on total throughput of D2D multicast groups is shown in Figure 4. It can be observed that the energy efficiency performance of our proposed scheme and the scheme in [14] both decrease with the increase of total throughput. This is explained as when the total throughput increases, according to our proposed power control scheme, the transmit power also increases, which contributes to decreased energy efficiency. Moreover, based on minimizing the interference of D2D clusters, our proposed heuristic algorithm allocates better cellular resource to D2D clusters. Meanwhile, it also decreases the transmit power of co-channel CUs, which contributes to improved EE performance of D2D multicast clusters.

Figure 5 plots the average SINR of D2D multicast clusters with different D2D cluster radius. We assume that there are totally 500 UEs randomly distributed in the simulation area. From the figure, we can infer that for the schemes which include power control scheme, such as our proposed algorithm and the scheme in [14], the SINR distribution does not obviously decrease with the increase of D2D group radius. This is because according to the changing channel conditions and resource reuse relationships, the power control method jointly adjusts the transmission power of both D2D clusters and CUs, which ensures reliable D2D multicast transmission [19]. On the contrary, for the algorithms which do not incorporate power control schemes, such as [7,18], the QoS requirement of D2D multicast clusters in terms of minimum SINR constraint cannot be guaranteed when the channel quality becomes worse.

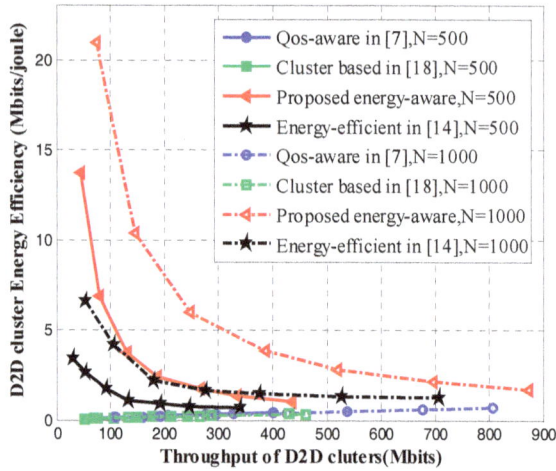

Figure 4. D2D Group Energy Efficiency versus throughput.

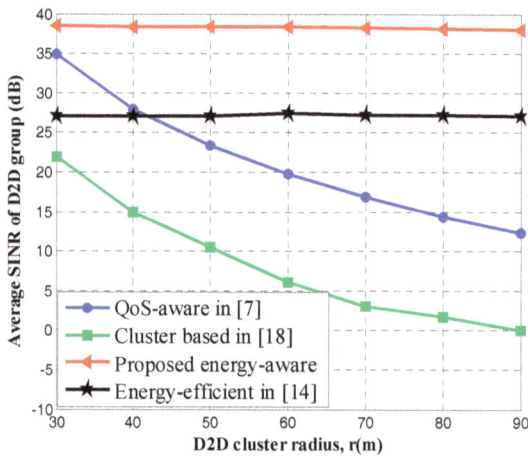

Figure 5. Average D2D group Energy efficiency versus number of group users.

Figure 6 plots the Cumulative Distribution Function (CDF) curve of cellular users' SINR concerning with different schemes. It can be obviously seen that with the introduction of power control scheme, the SINR curves of both [14] and our proposed scheme decrease, while the SINR curves of [7,18] are not severely impacted. This is because our proposed strategy decreases the transmit power of CUs in order to maximize the total energy-efficiency of D2D multicast clusters, which results in deterioration of CU's SINR. However, compared with [14], our resource allocation strategy will pick up a CU who brings the least interference for the D2D multicast cluster. Hence, the co-channel interference caused by underlay D2D transmission will be effectively controlled, which contributes to improved SINR of C-links. From Figure 6, we can also infer that although the link quality of CUs has been affected a little for the proposed scheme. But the performance degradation is still acceptable to CUs because the power allocation strategy considers the minimum SINR constraint of different users.

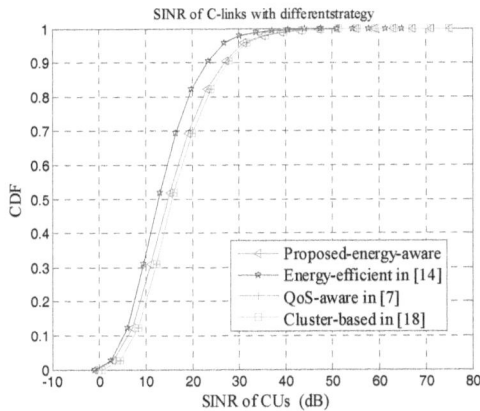

Figure 6. Signal-to-interference-and-noise ratio (SINR) of C-links with different schemes.

6. Conclusions

In this paper, we have investigated an energy-efficient resource and power allocation algorithm for multicast D2D communication underlying a cellular network. The goal of this paper is to maximize the overall energy efficiency of D2D multicast groups through appropriate resource allocation and a power control scheme, while maintaining the SINR requirements of both CUs and D2D clusters. A heuristic resource and power control algorithm is then proposed to solve the above problem with less complexity. It is shown by simulation that the energy efficiency of D2D multicast clusters with the proposed scheme can be improved significantly compared with conventional resource allocation schemes. In future, we plan to extend this work to the multiple multicast problem which means that several multicast clusters are allowed to reuse the same RB of a CU to further improve the spectrum efficiency.

Acknowledgments: This work is supported by the National Natural Science Foundation of China (No. 61501371), National Science and Technology Major Project of the Ministry of Science and Technology of China(project number: 2017ZX03001012-005). The international Exchange and Cooperation Projects of Shaanxi Province (project number: 2016KW-046).

Author Contributions: Fan Jiang conceived and wrote the paper, Honglin Wang and Hao Ren designed and conducted comprehensive computer simulations, Shuai Xu helped Fan Jiang to conduct the theory analysis.

Conflicts of Interest: The authors declare no conflict of interest.

References

1. IMT-2020 (5G) Promotion Group. White Paper on 5G Wireless Technology Architecture. Available online: http://www.imt-2020.cn/zh (accessed on 15 May 2015).
2. Tehrani, M.N.; Uysal, M.; Yanikomeroglu, H. Device-to-device communication in 5G cellular networks: Challenges, solutions and future directions. *IEEE Commun. Mag.* **2014**, *52*, 86–92. [CrossRef]
3. Wei, L.L.; Hu, R.Q.; Qian, Y. Enable device-to-device communications underlying cellular networks: Challenges and research aspects. *IEEE Commun. Mag.* **2014**, *52*, 90–96. [CrossRef]
4. Condoluci, M.; Militano, L.; Araniti, G. Multicasting in LTE-A networks enhanced by device-to-device communications. In Proceedings of the 2013 IEEE Globecom Workshops (GC Workshops), Atlanta, GA, USA, 9–13 December 2013; pp. 567–572.
5. Shang, B.; Zhao, L.; Chen, K.C. Operator's Economy of Device-to-Device Offloading in Underlaying Cellular Networks. *IEEE Commun. Lett.* **2017**, *21*, 865–868. [CrossRef]

6. Meshgi, H.; Zhao, D.; Zheng, R. Joint channel and power allocation in underlay multicast device-to-device communications. In Proceedings of the 2015 IEEE International Conference on Communications (ICC), London, UK, 8–12 June 2015; pp. 2937–2942.

7. Bhardwaj, A.; Agnihotri, S. A resource allocation scheme for device-to-device multicast in cellular networks. In Proceedings of the 2015 IEEE 26th Annual International Symposium on Personal, Indoor and Mobile Radio Communications (PIMRC), Hong Kong, China, 30 August–2 September 2015; pp. 1498–1502.

8. Kitagawa, K.; Homma, H.; Suegara, Y. A user selection algorithm for D2D multicast communication underlying cellular systems. In Proceedings of the 2017 IEEE Wireless Communications and Networking Conference (WCNC 2017), San Francisco, CA, USA, 19–22 March 2017; pp. 1–6.

9. Yu, S.; Langar, R.; Wang, X. A D2D-Multicast Based Computation Offloading Framework for Interactive Applications. In Proceedings of the 2016 IEEE Global Telecommunications Conference (GLOBECOM 2016), Washington, DC, USA, 4–8 December 2016; pp. 1–6.

10. Zhao, P.; Feng, L.; Yu, P. Resource allocation for energy-efficient Device-to-Device multicast communication. In Proceedings of the 2016 19th International Symposium on Wireless Personal Multimedia Communications (WPMC 2016), Shenzhen, China, 14–16 November 2016; pp. 518–523.

11. Chen, X.; Pu, L.; Gao, L. Exploiting massive D2D collaboration for energy-efficient mobile edge computing. *IEEE Wirel. Commun.* **2016**, *24*, 64–71. [CrossRef]

12. Jiang, L.; Tian, H.; Xing, Z. Social-aware energy harvesting device-to-device communications in 5G networks. *IEEE Wirel. Commun.* **2016**, *23*, 20–27. [CrossRef]

13. Xu, L.; Jiang, C.; Shen, Y. Energy efficient D2D communications: A perspective of mechanism design. *IEEE Trans. Wirel. Commun.* **2016**, *15*, 7272–7285. [CrossRef]

14. Guan, N.; Zhou, Y.; Liu, H. An energy efficient cooperative multicast transmission scheme with power control. In Proceedings of the 2011 IEEE Global Telecommunications Conference (GLOBECOM 2011), Houston, TX, USA, 5–9 December 2011; pp. 1–5.

15. Feng, D.; Yu, G.; Xiong, C. Mode switching for energy-efficient device-to-device communications in cellular networks. *IEEE Trans. Wirel. Commun.* **2015**, *14*, 6993–7003. [CrossRef]

16. Feng, D.Q.; Lu, L.; Wu, Y.Y.; Li, G.Y.; Li, S.Q. Device-to-device communications underlying cellular Networks. *IEEE Trans. Commun.* **2013**, *61*, 3541–3551. [CrossRef]

17. Peng, B.; Peng, T.; Liu, Z. Cluster-based multicast transmission for Device-to-Device (D2D) communication. In Proceedings of the 2013 IEEE 78th Vehicular Technology Conference (VTC Fall), Las Vegas, NV, USA, 2–5 September 2013; pp. 1–5.

18. Zhang, G.; Yang, K.; Chen, H.H. Socially aware cluster formation and radio resource allocation in D2D networks. *IEEE Wirel. Commun.* **2016**, *23*, 68–73. [CrossRef]

19. Ren, Y.; Liu, F.; Liu, Z.; Ji, Y. Power control in D2D-based vehicular communication networks. *IEEE Trans. Veh. Technol.* **2015**, *64*, 5547–5562. [CrossRef]

future internet

MDPI

Article

A Practical Resource Management Scheme for Cellular Underlaid D2D Networks

Tae-Won Ban [†]

Department of information and communication engineering, Gyeongsang National University, Tongyeong-si 46764, Korea; twban35@gnu.ac.kr; Tel.: +82-55-772-9177
† Current address: Marine Science Bldg. 807, Tongyeong-si 46764, Korea.

Received: 27 September 2017; Accepted: 8 October 2017; Published: 13 October 2017

Abstract: In this paper, we investigate a resource management scheme for cellular underlaid device-to-device (D2D) communications, which are an integral part of mobile caching networks. D2D communications are allowed to share radio resources with cellular communications as long as the generating interference of D2D communications satisfies an interference constraint to secure cellular communications. Contrary to most of the other studies, we propose a distributed resource management scheme for cellular underlaid D2D communications focusing on a practical feasibility. In the proposed scheme, the feedback of channel information is not required because all D2D transmitters use a fixed transmit power and every D2D transmitter determines when to transmit data on its own without centralized control. We analyze the average sum-rates to evaluate the proposed scheme and compare them with optimal values, which can be achieved when a central controller has the perfect entire channel information and the full control of all D2D communications. Our numerical results show that the average sum-rates of the proposed scheme approach the optimal values in low or high signal-to-noise power ratio (SNR) regions. In particular, the proposed scheme achieves almost optimal average sum-rates in the entire SNR values in practical environments.

Keywords: D2D; cellular-aided D2D; underlay; mobile caching

---

1. Introduction

Mobile internet traffic has been explosively increasing in recent years [1]. To be specific, multimedia video traffic accounts for about 60% of total mobile internet traffic and the ratio is expected to grow to 78% by 2021 [1]. The next generation mobile communication systems requires a much higher capacity to support the explosively increasing multimedia data. It is the easiest way to increase capacity to use wider bandwidth, but radio spectrum, unfortunately, is a limited resource. Many promising technologies such as multiple input and multiple output (MIMO) and small cell systems have been investigated to enhance the spectral efficiency. However, the spectral efficiency is affected by radio channels, which are mainly determined by the distance between transmitters and receivers. This is the reason that higher order modulations can be only applied to devices near base stations (BSs) in current communication systems such as long-term evolution (LTE) and wireless local area network (WLAN). Thus, it is the most effective way to increase the spectral efficiency to reduce the distance between transmitters and receivers rather than other promising technologies.

On the other hand, the quality of service (QoS) for multimedia services is mainly determined by not only transmission rate but also latency, and the latency is closely related to the physical distance between clients and content servers. No matter how much we increase the transmission rate, we can not reduce the latency below a certain level because multimedia data is currently transferred from a content server to mobile clients through many intermediate network entities. If we can shorten the physical distance in end-to-end communications, both the spectral efficiency and the latency will be greatly enhanced at the same time [2]. The communication distance can be noticeably reduced by using mobile caching technologies, where multimedia data is cached in mobile devices and thus can be directly transferred to other mobile devices without going through intermediate nodes [3–6]. Device-to-device (D2D) communication is one of the integral parts for the mobile caching networks. Motivated by these contexts, D2D communications have been attracting plenty of interest as one of the promising technologies for the next generation mobile communications systems [7–17]. Furthermore, we can achieve much higher spectral efficiency by cellular underlaid D2D networks, where D2D communications share radio resources with conventional cellular communications as long as the generating interference of D2D communications is regulated to secure cellular communications [9,10].

Despite the extensive previous research on D2D communications, the practical feasibility of cellular underlaid D2D communication networks is not guaranteed because most of them require immoderate intervention of cellular infrastructure such as BS or excessive signalling overhead for channel information feedback. In this paper, we thus investigate a cellular underlaid D2D communication network by focusing on the practical feasibility and we propose a practically feasible resource management scheme for cellular underlaid D2D networks. In the proposed scheme, all D2D transmitters use a fixed transmit power level to remove the signalling for channel information feedback, and each D2D transmitter can determine whether to transmit data on its own without explicit control from BS, while satisfying the interference constraint imposed by cellular networks. The performance of the proposed scheme is evaluated in terms of average sum-rates and the feasibility is also verified in practical environments.

The rest of this paper is organized as follows. In Section 3, system and channel models are described. In Section 4, a practical resource management scheme is proposed. Our numerical results are shown in Section 5. Finally, this paper is concluded in Section 6.

2. Related Work

The different transmission modes were introduced for cellular D2D communications [11]: (1) non-orthogonal mode where D2D communications share the same resource as cellular communications, (2) orthogonal mode where D2D communications use dedicated resources, and (3) cellular mode where D2D traffic is transferred through a BS. Based on the three modes, an optimal mode selection scheme was proposed to enhance the performance of D2D communications. While single transmit and receiver antenna are considered in [11], multiple input and multiple output (MIMO) was taken into account by designing pre-coding matrices at each node [12]. In [13], a non-orthogonal centralized D2D communication system was investigated. It was assumed that D2D communications share uplink frequency with cellular communications, as in this paper, and each device can transmit its data in D2D or cellular mode via a BS. An optimal mode selection scheme was proposed to maximize the overall sum-rate. However, the proposed scheme is centralized and can not guarantee the QoS of cellular communications, unlike the proposed scheme in this paper. On the other hand, there have been many studies to investigate power control schemes to deal with cross interference between D2D and cellular communications, which is one of the challenging problems to limit the performance of cellular underlaid D2D communications [14–17]. Optimization problems have been formulated in both non-orthogonal and orthogonal modes and proposed optimal transmit power allocation schemes to maximize the effective capacity based on the formulated optimization problems, while satisfying different delay–QoS requirements [14]. Several suboptimal power control schemes were also proposed and the performance was analysed by simulations. Despite their excellent performance, the power control schemes inevitably yield excessive complexity and signalling overhead. Thus, a distributed power control algorithm was proposed in [15]. The algorithm simply aims to set the individual signal-to-noise and interference ratio (SINR) targets such that the required sum power is minimized with respect to a sum rate target and allocates transmit power levels. MIMO was also considered. However, the simple distributed algorithm can not guarantee the overall performance such as sum-rate and QoS of cellular communications. Contrary to most of the studies considering the uplink of cellular networks, the impact of D2D communication on the downlink coverage of a cellular network was investigated in [16]. They developed an analytical model to characterize the coverage probability of cellular networks where a D2D link exists. Shadowing and power control were considered, and BSs and devices were distributed by a Poisson point process. Both centralized and distributed power control schemes were also proposed to maximize the performance of D2D links in terms of sum-rate [17]. However, the QoS of cellular communications was not strictly guaranteed despite their priority over D2D communications.

In spite of many previous studies, to the best of our knowledge, there have been no studies that can strictly guarantee the QoS of cellular communications and provide a moderate level of complexity and signalling overhead for commercialization, which motivated the study in this paper.

3. System and Channel Models

In this paper, we investigate a cellular underlaid D2D wireless communication network as depicted in Figure 1. The utilization in cellular uplink is much lower than in downlink because of the asymmetry of mobile internet traffic, and thus it is more efficient for the underlaid D2D network to share uplink resources with cellular communications [18,19]. However, the resource sharing between D2D and cellular communications can cause the quality deterioration in conventional cellular communications, which have higher priority over D2D communications. Thus, a cellular infrastructure such as BS imposes an interference constraint on D2D communications to secure the quality of cellular communications. In our system model, we have N D2D pairs, a cellular BS, and a cellular user equipment (UE). In this paper, the association process of D2D pairs and cellular uplink resource scheduling are both beyond the scope of the paper. Thus, we assume that each D2D pair has been already associated. A D2D receiver that wants to receive content is associated with one of the D2D transmitters that cache the wanted content and are located in the proximity of the receiver. In addition, a cellular UE is scheduled to transmit its uplink data to the cellular BS based on a scheduling policy. In this paper, we use a statistical channel model, where g_{ij} denotes the channel coefficient between transmitter i and receiver j. $i \in \{1, 2, \cdots, N\}$ or $i = u$ and $j \in \{1, 2, \cdots, N\}$ or $j = b$. $i = u$ denotes a cellular UE and $j = b$ denotes a cellular BS. We consider a Rayleigh fading model, and thus $|g_{ij}|^2$ is exponentially distributed with mean value λ_{ij} [20]. The effect of path loss can be incorporated into λ_{ij}. Quasi-static block fading is also considered and thus all the channel coefficients are constant during one frame for data transmission and randomly vary each frame. We assume that D2D communications adopt a time division duplex (TDD) scheme and thus the D2D channels are reciprocal without loss of generality. N_0 denotes the variance of additive white Gaussian noise (AWGN) in D2D receivers. We assume that the transmit power levels of all D2D transmitters and cellular transmitter are fixed at P. The cellular transmitter can immediately transmit its data regardless of the presence of D2D communications, while D2D transmissions should be controlled to protect the quality of cellular uplink communications. Thus, a cellular BS imposes an interference constraint on D2D communications. That is, the total interference power that all D2D transmissions cause to the cellular BS should be less than I_{th} at any moment. I_{th} is a parameter that the cellular BS can determine by considering both the quality of cellular communications and the performance of D2D communications. Because D2D transmitters use a fixed transmit power, $P_i = P$ if D2D transmitter i is allowed to transmit its data; otherwise, $P_i = 0$. Then, the total interference power received at the cellular BS from all D2D transmitters can be calculated as $\sum_{i=1}^{N} I_i P |g_{ib}|^2$, where I_i is an indicator function to denote the activity of D2D transmitter i, defined as

$$I_i = \begin{cases} 1, & \text{if D2D transmitter } i \text{ transmits data } (P_i = P), \\ 0, & \text{otherwise } (P_i = 0). \end{cases} \qquad (1)$$

If we define $\mathcal{I} \triangleq \{I_1, \cdots, I_N\}$, D2D communications should comply with the following interference constraint imposed by the BS;

$$\sum_{i=1}^{N} I_i P |g_{ib}|^2 \leq I_{th} \tag{2}$$

to protect the cellular uplink communications securely. If normalized by N_0, Equation (2) can be rewritten as

$$\sum_{i=1}^{N} I_i \rho |g_{ib}|^2 \leq \frac{I_{th}}{N_0} \triangleq I'_{th}, \tag{3}$$

where ρ is a transmit power-to-noise ratio of D2D transmitters, hereafter simply called SNR. Then, the signal to noise plus interference (SINR) received at a D2D receiver i, denoted by γ_i, can be calculated as

$$\begin{aligned}
\gamma_i(\mathcal{I}) &= \frac{I_i P |g_{ii}|^2}{\sum_{k=1, k \neq i}^{N} I_k P |g_{ki}|^2 + P |g_{bi}|^2 + N_0} \\
&= \frac{I_i \rho |g_{ii}|^2}{\sum_{k=1, k \neq i}^{N} I_k \rho |g_{ki}|^2 + \rho |g_{bi}|^2 + 1}, \tag{4}
\end{aligned}$$

for a given $\mathcal{I}$. Then, the sum-rate of a cellular underlaid D2D network can be calculated by $\sum_{i=1}^{N} \log_2 (1 + \gamma_i(\mathcal{I}))$. Finally, we need to choose an optimal $\mathcal{I}$ for maximizing the sum-rate while satisfying the interference constraint as follows:

$$\mathcal{I}^* = \arg\max_{\mathcal{I}} \sum_{i=1}^{N} \log_2 (1 + \gamma_i(\mathcal{I}))$$

$$\text{s.t.}$$

$$\sum_{i=1}^{N} I_i \rho |g_{ib}|^2 \leq I'_{th}. \tag{5}$$

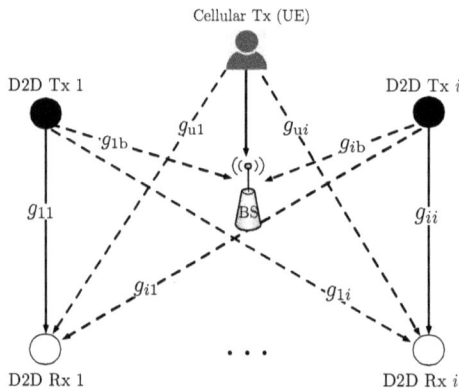

Figure 1. An underlaid device-to-device (D2D) communication network.

4. Proposed Resource Management for Cellular Underlaid D2D Networks

The optimization in Equation (5) requires tremendous complexity because it is not a convex problem but a combinatorial problem. In addition, a central node such as a cellular BS should have perfect information of entire channels to solve the problem and have full control of D2D communications, which inevitably causes excessive signalling overhead for the feedback of channel state information (CSI). In particular, both the complexity and signalling overhead exponentially increase as the number of D2D pairs N increases. In this paper, we thus propose a fully distributed resource management scheme for cellular underlaid D2D communication networks. The proposed scheme does not require any signalling overhead for CSI and can be operated in a fully distributed manner with a simple signalling broadcast by the BS. Figure 2 shows the flow diagram for our proposed scheme.

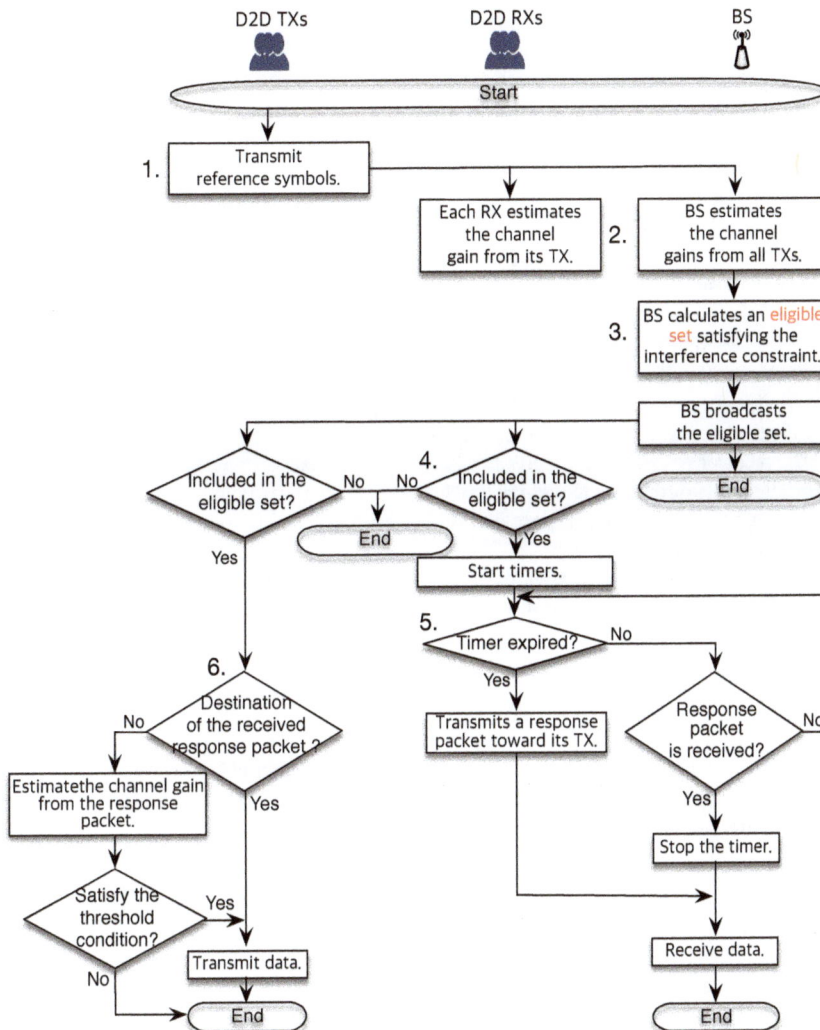

Figure 2. Flow diagram of the proposed scheme for cellular underlaid D2D communication networks.

1. Every D2D transmitter transmits reference symbols to enable D2D receivers and BS to estimate channel gains. We assume that each D2D transmitter is assigned orthogonal radio resource for the reference symbols.

2. Each D2D receiver only estimates the channel gain received from its associated transmitter, while the BS estimates all channel gains received from all N D2D transmitters.

3. The BS sorts N measured channel gains in an ascending order. The D2D transmitter with the i-th largest channel gain is indexed by $\hat{i}$. Then, the sorted set of D2D transmitters can be denoted by $\{\hat{1}, \cdots, \hat{i}, \cdots, \hat{N}\}$ and $|g_{\hat{1}b}|^2 \leq \cdots \leq |g_{\hat{i}b}|^2 \cdots \leq |g_{\hat{N}b}|^2$ is satisfied. The BS determines an eligible set $\mathcal{E}$, defined by

$$\mathcal{E} = \{\hat{1}, \hat{2}, \cdots, \hat{i}^*\},\tag{6}$$

where $\hat{i}^*$ can be obtained by

$$\hat{i}^* = \max\left\{\hat{i} \,\middle|\, \sum_{k=\hat{1}}^{\hat{i}} |g_{\hat{k}b}|^2 \leq I_{th}\right\}.\tag{7}$$

Even if all D2D transmitters in the eligible set transmit data simultaneously, the interference constraint imposed by BS is satisfied. Thus, no matter which combination of D2D transmitters in the eligible set transmit data, the interference constraint will be satisfied.

The BS broadcasts a bitmap message with N bits to notify which D2D transmitters are included in the eligible set $\mathcal{E}$.

4. If each D2D transmitter $\hat{i}$ is included in the eligible set, then it waits for a response packet, which will be transmitted from a D2D receiver in Step 5. Otherwise, it terminates this algorithm. If each D2D receiver $\hat{i}$ is included in the eligible set, then it starts a timer. Otherwise, it terminates this algorithm. The value of timer is determined by $\frac{C}{|g_{\hat{i}\hat{i}}|^2}$, where C is a constant to be determined by a controller and $|g_{\hat{i}\hat{i}}|^2$ was measured in Step 2.

5. A D2D receiver whose timer expires first, denoted by $\hat{t}$, transmits a response packet towards its associated transmitter $\hat{t}$. Note that the timer of the D2D receiver that has the largest channel gain will expire first and thus a D2D pair with the best channel condition can be selected satisfying the interference constraint.

6. All D2D transmitters can receive the response packet transmitted by the receiver $\hat{t}$ in Step 5. The D2D transmitter $\hat{t}$ can transmit its data, but each D2D transmitter $\hat{i}$ included in $\mathcal{E} \setminus \{\hat{t}\}$ checks the following condition:

$$\frac{|g_{\hat{i}\hat{i}}|^2}{|g_{\hat{i}\hat{t}}|^2} \geq \eta, \ \hat{i} \in \mathcal{E} \setminus \{\hat{t}\},\tag{8}$$

where $|g_{\hat{i}\hat{t}}|^2$ can be obtained by measuring the response packet transmitted by receiver $\hat{t}$, and η is a threshold value required in the proposed scheme. All other D2D receivers except $\hat{t}$ stop their timers right after receiving the response packet from $\hat{t}$.

The set of D2D transmitters to transmit data simultaneously can be determined by $\mathcal{T} \triangleq \{\hat{t}\} \cup \left\{\hat{i} \,\middle|\, \frac{|g_{\hat{i}\hat{i}}|^2}{|g_{\hat{i}\hat{t}}|^2} \geq \eta, \ \hat{i} \in \mathcal{E} \setminus \{\hat{t}\}\right\}$, satisfying the interference constraint. The D2D pairs with higher channel gain and lower generating interference to the receiver $\hat{t}$ are more likely to be selected to transmit data along with the D2D pair $\hat{t}$. In summary, the proposed scheme is based on a well-known fact that opportunistic resource management schemes can greatly reduce the complexity [21]. Moreover, the proposed scheme selects one user with the highest channel gain out of an eligible set and extra users out of the eligible set satisfying Equation (8) in a decentralized manner to remove most of the complexity and feedback overhead. For a given threshold η, the SINR of a D2D pair i in $\mathcal{T}$ can be calculated as

$$\gamma_{i\in\mathcal{T}}^{\text{prop}}(\eta) = \frac{\rho|g_{ii}|^2}{\sum_{k\in(\mathcal{T}\setminus\{i\})} \rho|g_{ki}|^2 + \rho|g_{bi}|^2 + 1},\tag{9}$$

and then the sum-rate of the proposed scheme can be obtained as $\sum_{i \in \mathcal{T}} \log_2 \left(1 + \gamma_i^{\text{prop}}(\eta)\right)$.

Finally, note that the optimal scheme should carry out 2^N iterations to calculate sum-rates with the entire channel information feedback at each frame and thus we can not afford the complexity as N increases, while the proposed scheme only requires an N-bit message that is broadcast by BS to inform an eligible set.

5. Numerical Results

In this section, we evaluate the performance of the proposed scheme in terms of average sum-rate and compare it with optimal performance. In addition, we derive the optimal threshold values to maximize the performance of the proposed scheme. Figure 3 shows the average sum-rates of the proposed scheme where all channels are assumed to be *i.i.d.* and thus $\lambda_{ij} = 1 \; \forall \; i$ and j. $N = 5$ and $I'_{\text{th}} = 0$ or 5 dB. For the comparison, it also shows the optimal average sum-rates obtained by solving Equation (5) based on the Brute Force searching algorithm. For a higher I'_{th} denoting that a cellular network tolerates higher interference from a D2D network, the cardinality for the eligible set of D2D transmitters increases and thus average sum-rates of the D2D network also increases due to an increasing gain of user selection diversity. As the SNR of D2D transmitters increases, the average sum-rates of the D2D network also increase due to transmit power gain. However, the SNR values higher than a moderate level reduce the average sum-rate on the contrary because the cardinality for the eligible set of D2D transmitters seriously decreases and thus the user selection diversity decreases as well. The proposed scheme can achieve almost optimal average sum-rates when SNR is low or high, while the average sum-rates of the proposed scheme is lower than the optimal value for moderate SNR values. It should be also noted that the proposed scheme can dramatically reduce the complexity and feedback overhead, compared to an optimal scheme. Figure 4 also shows average sum-rates of the proposed scheme and optimal sum-rates under the same conditions as in Figure 3, except for $N = 10$. As N increases, the gain of user selection diversity increases and thus the overall performance is enhanced. The average sum-rates of the proposed and optimal schemes are tabulated and the ratios of average sum-rate obtained by the proposed scheme to optimal value are also summarized in Table 1.

Figures 5 and 6 show the optimal threshold values to maximize the average sum-rates of the proposed scheme with the same parameter values with Figures 3 and 4, respectively. As the SNR of D2D transmitters increases to a moderate level, the optimal threshold value increases and thus $|\mathcal{T}|$ decreases. The optimal threshold value of the proposed scheme decreases if the SNR increases above the moderate level. The excessively high SNR above the moderate level seriously decrease $|\mathcal{E}|$ and thus a low threshold can enhance the performance of the proposed scheme by increasing $|\mathcal{T}|$.

In Figures 7 and 8, we analyze the average sum-rates of the proposed scheme when channels are non-*i.i.d.* to verify the feasibility of the proposed scheme in practical environments. Table 2 summarizes the average sum-rates of the proposed and optimal schemes and the ratios of average sum-rate obtained by the proposed scheme to optimal value. $(2 \times N)$ D2D nodes are uniformly distributed in a circle with a radius of 200 m and a cellular BS is located at the center of the circle. The average channel gain between transmitter i and receiver j is determined by $\lambda_{ij} = \min\left(-30 \text{ dB}, L_{ij}^{-4}\right)$, where -30 dB and $L_{ij}(0 \leq L_{ij} \leq 200)$ denote a minimum coupling loss and the distance between transmitter i and receiver j, respectively, and $i \in \{1, 2, \cdots, N, u\}, j \in \{1, 2, \cdots, N, b\}$. It is assumed that the channel bandwidth is 10 MHz and thermal noise power density is -174 dBm/Hz. The transmit power of all transmitters varies from 0–30 dBm and $I_{\text{th}} = -50$ or -80 dBm. The performance gap between the proposed and optimal schemes is noticeably reduced in the entire SNR region, compared to the *i.i.d.* channel environments. It is shown that the average sum-rates of the proposed scheme approach the optimal values for entire SNR values and they are almost optimal in a practical region of SNR such as 0–20 dBm.

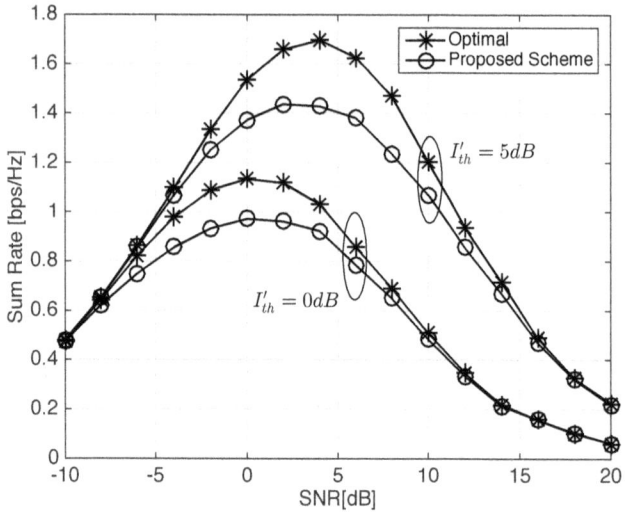

Figure 3. Average sum-rates when all channels are *i.i.d.*, $N = 5$, and $I'_{\text{th}} = 0$ or 5 dB.

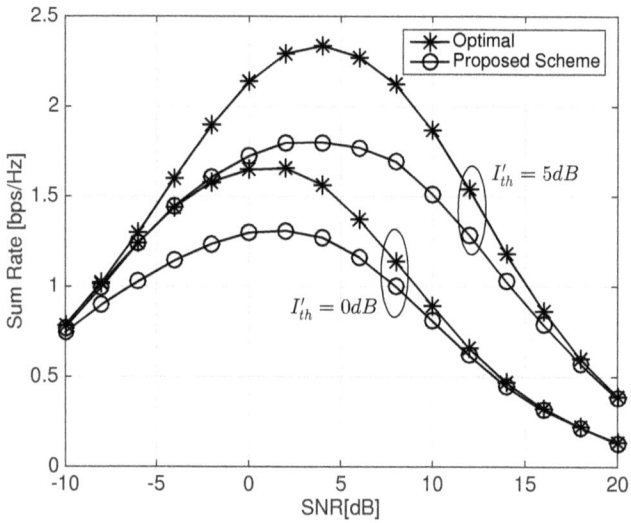

Figure 4. Average sum-rates when all channels are *i.i.d.*, $N = 10$, and $I'_{\text{th}} = 0$ or 5 dB.

Table 1. Average sum-rates when all channels are *i.i.d.*

SNR (dB)	N = 5						N = 10					
	I'_{th} = 0 dB			I'_{th} = 5 dB			I'_{th} = 0 dB			I'_{th} = 5 dB		
	Prop	Opt	Ratio	Prop	Opt	Ratio	Prop	Opt	Ratio	Prop	Opt	Ratio
−10	0.47	0.48	99.5%	0.48	0.48	99.8%	0.75	0.79	94.7%	0.77	0.78	98.3%
−6	0.75	0.82	91.3%	0.86	0.86	99.1%	1.03	1.24	82.8%	1.24	1.30	95.3%
−2	0.93	1.09	85.5%	1.25	1.34	93.8%	1.24	1.58	78.2%	1.61	1.90	84.8%
2	0.96	1.12	86.0%	1.44	1.66	86.6%	1.31	1.66	79.1%	1.80	2.29	78.5%
6	0.78	0.86	91.6%	1.38	1.63	84.9%	1.16	1.37	84.8%	1.77	2.27	77.9%
10	0.48	0.51	94.8%	1.07	1.20	89.0%	0.81	0.89	90.7%	1.52	1.87	81.1%
14	0.21	0.22	98.1%	0.67	0.72	93.6%	0.45	0.47	95.4%	1.03	1.18	86.9%
18	0.10	0.10	97.8%	0.32	0.33	98.6%	0.22	0.22	99.8%	0.57	0.60	95.5%

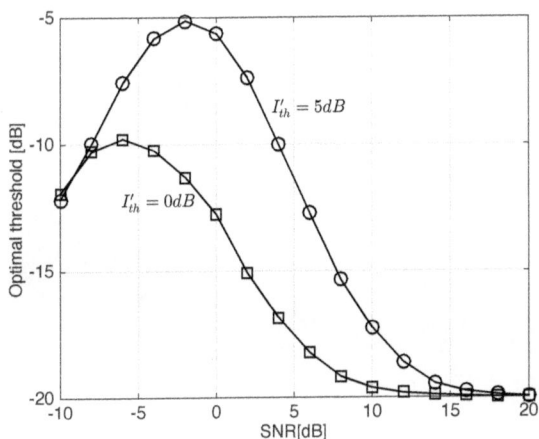

Figure 5. Optimal threshold in the proposed scheme when all channels are *i.i.d.*, $N = 5$, and $I'_{th} = 0$ or 5 dB.

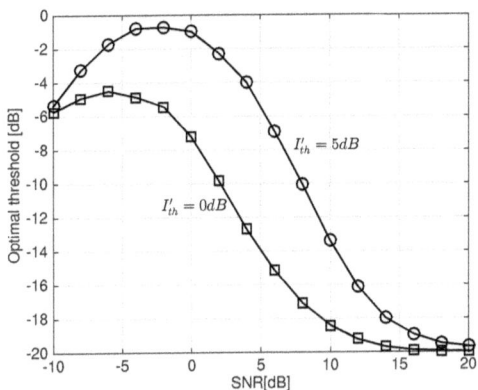

Figure 6. Optimal threshold in the proposed scheme when all channels are *i.i.d.*, $N = 10$, and $I'_{th} = 0$ or 5 dB.

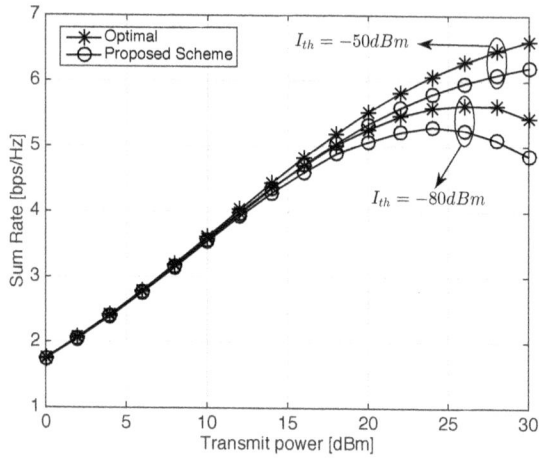

Figure 7. Average sum-rates when all channels are non-*i.i.d.* and $N = 5$.

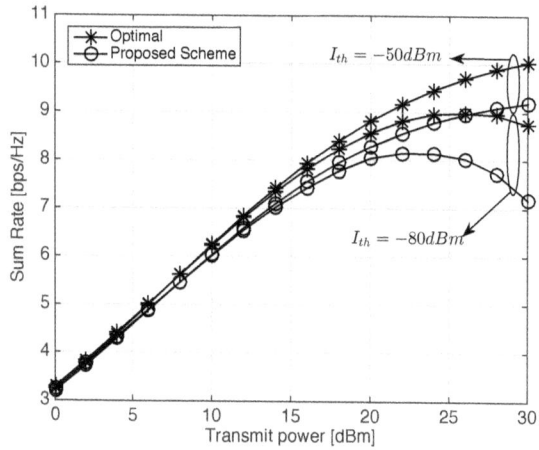

Figure 8. Average sum-rates when all channels are non-*i.i.d.* and $N = 10$.

Table 2. Average sum-rates when all channels are non-*i.i.d.*

P (dBm)	N = 5						N = 10					
	$I_{th} = -80$ dBm			$I_{th} = -50$ dBm			$I_{th} = -80$ dBm			$I_{th} = -50$ dBm		
	Prop	Opt	Ratio	Prop	Opt	Ratio	Prop	Opt	Ratio	Prop	Opt	Ratio
0	1.74	1.75	99.7%	1.75	1.75	99.7%	3.25	3.29	98.7%	3.20	3.24	98.8%
4	2.39	2.41	99.4%	2.41	2.42	99.4%	4.32	4.41	98.0%	4.29	4.38	98.1%
8	3.14	3.17	99.1%	3.17	3.20	99.0%	5.46	5.62	97.3%	5.46	5.62	97.2%
12	3.92	3.98	98.5%	3.96	4.03	98.4%	6.54	6.80	96.2%	6.59	6.85	96.2%
16	4.59	4.70	97.8%	4.71	4.83	97.6%	7.42	7.82	94.9%	7.55	7.94	95.2%
20	5.07	5.26	96.4%	5.34	5.52	96.6%	8.05	8.57	93.9%	8.29	8.82	94.0%
24	5.30	5.59	94.8%	5.80	6.07	95.5%	8.15	8.97	90.8%	8.79	9.46	92.9%
28	5.10	5.62	90.7%	6.10	6.46	94.4%	7.73	8.95	86.4%	9.09	9.89	92.0%

6. Conclusions

In this paper, we investigated a cellular underlaid D2D network, which is an enabling technology for mobile caching services. D2D communications are allowed to share radio resources with cellular uplink communications under the assumption that they comply with an interference constraint imposed by a cellular controller such as BS to secure the quality of cellular communications. The performance of D2D communications can be maximized by selecting an optimal combination of D2D pairs to transmit data. However, it causes tremendous complexity of computations and signalling overhead for channel feedbacks to determine an optimal combination complying with the interference constraint. Thus, we proposed a practical resource management scheme for D2D communications. Each D2D pair determines whether to transmit data on its own based on a threshold value and simple bitmap information broadcast by the BS. Thus, the proposed scheme does not require any feedback from the D2D network and does not cause any computational complexity to a BS either. We evaluated the performance of the proposed scheme in terms of average sum-rate and compared it with the optimal scheme. We also derived the optimal threshold values that maximize the average sum-rate of the proposed scheme. Our numerical results showed that the average sum-rates of the proposed scheme approach the optimal sum-rates in low or high SNR regions, despite the tremendous reduction in complexity and signalling overhead. It was also shown that the gap in performance between the proposed and optimal schemes noticeably decreases in the entire SNR region in practical environments where channels are non-*i.i.d.*

Acknowledgments: This work was supported by the Gyeongsang National University Fund for Professors on Sabbatical Leave, 2017 and Institute for Information & communications Technology Promotion(IITP) grant funded by the Korea government(MSIT) (No.2015-0-00820, A research on a novel communication system using storage as wireless communication resource).

Conflicts of Interest: The author declares no conflict of interest.

References

1. Cisco, Visual Networking Index: Global Mobile Data Traffic Forecast Update, 2016–2021, Whitepaper, 2017. Available online: https://www.cisco.com/c/en/us/solutions/collateral/service-provider/visual-networking-index-vni/mobile-white-paper-c11-520862.html (accessed on 1 October 2017).
2. Caire, G. The Role of Caching in 5G Wireless Networks, Invited Talk in IEEE ICC 2013. Available online: http://icc2013.ieee-icc.org/3_caire_icc-invited-talk-2013.pdf (accessed on 1 October 2017).
3. Cai, X.; Zhang, S.; Zhang, Y. Economic analysis of cache location in mobile network. In Proceedings of the IEEE WCNC, Shanghai, China, 7–10 April 2013; pp. 1243–1248.
4. Wang, X.; Chen, M.; Taleb, T.; Ksentini, A.; Leung, V.C.M. Cache in the air: Exploiting content caching and delivery techniques for 5G systems. *IEEE Commun. Mag.* **2014**, *52*, 131–139.
5. Chae, S.H.; Choi, W. Caching placement in stochastic wireless caching helper networks: Channel selection diversity via caching. *IEEE Trans. Wirel. Commun.* **2016**, *15*, 6626–6637.
6. Hong, B.; Choi, W. Optimal storage allocation for wireless cloud caching systems with a limited sum storage capacity. *IEEE Trans. Wirel. Commun.* **2016**, *15*, 6010–6021.
7. Giordano, D.; Traversol, S.; Grimaudo, L.; Mellia, M.; Baralis, E.; Tongaonkar, A.; Saha, S. YouLighter: A cognitive approach to unveil YouTube CDN and changes. *IEEE Trans. Cogn. Commun. Netw.* **2015**, *1*, 161–174.
8. Son, J. Content Networking Trends: OTT, Global CDN and Operator. Available online: http://www.netmanias.com/en/post/reports/6015/cdn-google-netflix-ott-transparent-cache-youtube/2013-content-networking-trends-ott-global-cdn-and-operator-cdn (accessed on 1 October 2017).
9. 3GPP TR 22.803 v12.2.0, Feasibility Study for Proximity Services (ProSe) (Release 12). Available online: http://www.tech-invite.com/3m22/tinv-3gpp-22-803.html (accessed on 1 October 2017).
10. 3GPP Work Programme. Available online: http://www.3gpp.org/DynaReport/GanttChart-Level-2.htm#bm580059 (accessed on 1 October 2017).

11. Yu, C.-H.; Doppler, K.; Ribeiro, C.; Tirkkonen, O. Resource Sharing Optimization for Device-to-Device Communication Underlaying Cellular Networks. *IEEE Trans. Wirel. Commun.* **2011**, *10*, 2752–2763.

12. Morattab, A.; Dziong, Z.; Sohraby, K.; Islam, M. An optimal MIMO mode selection method for D2D transmission in cellular networks. In Proceedings of the IEEE 11th International Conference on Wireless and Mobile Computing, Networking and Communications (WiMob), Abu Dhabi, UAE, 19–21 October 2015; pp. 392–398.

13. Hakola, S.; Chen, T.; Lehtomaki, J.; Koskela, T. Device-To-Device (D2D) Communication in Cellular Network—Performance Analysis of Optimum and Practical Communication Mode Selection. In Proceedings of the IEEE Wireless Communications and Networking Conference (WCNC), Sydney, Australia, 18–21 April 2010; pp. 1–6.

14. Cheng, W.; Zhang, X.; Zhang, H. Optimal Power Allocation With Statistical QoS Provisioning for D2D and Cellular Communications Over Underlaying Wireless Networks. *IEEE J. Sel. Areas Commun.* **2016**, *34*, 151–162.

15. Fodor, G.; Reider, N. A Distributed Power Control Scheme for Cellular Network Assisted D2D Communications. In Proceedings of the IEEE Global Telecommunications Conference (GLOBECOM), Houston, TX, USA, 5–9 December 2011; pp. 1–6.

16. Al-Rimawi, A.; Dardari, D. Analytical modeling of D2D communications over cellular networks. In Proceedings of the IEEE International Conference on Communications (ICC), London, UK, 8–12 June 2015; pp. 2117–2122.

17. Lee, N.; Lin, X.; Andrews, J.; Heath, R. Power Control for D2D Underlaid Cellular Networks: Modeling, Algorithms, and Analysis. *IEEE J. Sel. Areas Commun.* **2015**, *33*, 1–13.

18. Wang, J.; Zhu, D.; Zhao, C.; Li, J.C.F.; Lei, M. Resource sharing of underlaying device-to-device and uplink cellular communications. *IEEE Commun. Lett.* **2013**, *17*, 1148–1151.

19. Ericsson, Frame Structure for D2D-Enabled LTE Carriers. In Proceedings of the Meeting: 3GPP R1-141387, Shenzhen, China, 31 March–4 April 2014.

20. Sklar, B. Rayleigh fading channels in mobile digital communication systems. I. Characterization. *IEEE Commun. Mag.* **1997**, *35*, 90–100.

21. Asadi, A.; Mancuso, V. A Survey on Opportunistic Scheduling in Wireless Communications. *IEEE Commun. Surv. Tutor.* **2013**, *15*, 1671–1688.

![future internet logo] *future internet*

MDPI

Article

Multicell Interference Management in Device to Device Underlay Cellular Networks

Georgios Katsinis [1], Eirini Eleni Tsiropoulou [2] and Symeon Papavassiliou [1,*]

[1] School of Electrical & Computer Engineering, National Technical University of Athens,
     9 Iroon Polytechniou str, Zografou, 15780 Athens, Greece; gkatsinis@netmode.ntua.gr
[2] Electrical and Computer Engineering Department, University of New Mexico,
     Albuquerque, NM 87131, USA; eirini@unm.edu
*   Correspondence: papavass@netmode.ntua.gr; Tel.: +30-210-772-2550

Received: 27 July 2017; Accepted: 3 August 2017; Published: 7 August 2018

Abstract: In this paper, the problem of interference mitigation in a multicell Device to Device (D2D) underlay cellular network is addressed. In this type of network architectures, cellular users and D2D users share common Resource Blocks (RBs). Though such paradigms allow potential increase in the number of supported users, the latter comes at the cost of interference increase that in turn calls for the design of efficient interference mitigation methodologies. To treat this problem efficiently, we propose a two step approach, where the first step concerns the efficient RB allocation to the users and the second one the transmission power allocation. Specifically, the RB allocation problem is formulated as a bilateral symmetric interaction game. This assures the existence of a Nash Equilibrium (NE) point of the game, while a distributed algorithm, which converges to it, is devised. The power allocation problem is formulated as a linear programming problem per RB, and the equivalency between this problem and the total power minimization problem is shown. Finally, the operational effectiveness of the proposed approach is evaluated via numerical simulations, while its superiority against state of the art approaches existing in the recent literature is shown in terms of increased number of supported users, interference reduction and power minimization.

Keywords: interference management; resource allocation; power control; multicell device to device underlay networks; game theory; linear programming

1. Introduction

Device to Device (D2D) communication has emerged as an add-on communication paradigm to the modern 5G wireless cellular networks [1]. In these networks, two types of users exist. The first type of users are the cellular users who communicate conventionally via the intervention of the evolved nodeB (eNB). The second type of users concerns the D2D users who are able to communicate directly with each other [2].

D2D communications can take place either overlaying or underlaying inband a cellular network or outband [3]. D2D users can act cooperatively, form clusters and facilitate content dissemination [4,5]. In the overlay case, D2D communications use dedicated resources, while in the underlay case D2D communications share common resources with the residual cellular network. In this paper, we focus on the underlay policy, where the cellular and the D2D users share common radio resources, organized into resource blocks (RBs). RBs are combinations of frequency and time symbols which are allocated to the users of the network [6]. All users of the network belong to a cell, where the central entity, i.e., the eNB, is in charge of the efficient RBs assignment to the users. RBs assignment and power control are key factors for the efficient operation of the network in terms of spectral and power efficiency. Though significant research efforts have been devoted to the RBs and power allocation in a single-cell environment [7–12], the impact and consideration of multi-cell environment in the case of D2D underlay cellular network has been considerably neglected.

In this paper, we aim at closing this gap in the literature and study the joint problem of RB and power allocation in a multicell D2D underlay cellular environment. Multiple neighbouring cells exist serving the users of the network. Every cell is assumed to use the same spectrum band with the other cells for reasons of spectrum economy. In every cell, both cellular and D2D users co-exist and share common RBs. D2D users may share RBs with several other D2D users and with at most one cellular user in a given cell. Though a RB can be used by only one cellular user in a cell, yet interference may occur either by the reuse of the same RB by other cellular users in neighbouring cells, or by several other pairs of D2D users reusing the same RB either in the same cell or neighbouring cells.

In this setting, efficient RBs assignment and power control are of fundamental importance. On one hand RBs reuse promotes spectrum economy, but on the other hand causes signal to interference and noise ratio (SINR) degradation to the co-sharing users. The users of the network establish connections to either the eNB or their D2D receiver and try to obtain a minimum quality of service (QoS) at their receiver. QoS is expressed at the physical layer via the SINR at the receiver of each link. Users may have multiple different requirements with respect to the SINR at their receiver.

The rest of the paper is structured as follows. Section 1.1 discusses the related work and Section 1.2 summarizes and introduces our contributions. Section 2 presents the system model adopted, while in Section 3 the problem under consideration is formally presented. Subsequently, Sections 4 and 5 present the first and second stage of the proposed approach respectively, where the problems of RB allocation and power control are described and solved. Finally, in Section 6 detailed numerical results are presented demonstrating the operational effectiveness and efficiency of the proposed methodology, while Section 7 concludes the paper.

1.1. Related Work

RB allocation approaches in the uplink of Long Term Evolution (LTE) networks have gained a lot of interest in the literature [13–15]. In this setting, the eNB needs to determine in an efficient and timely manner the assignment of RBs to the users of the network. The optimization objective of the RBs assignment process could be the total throughput maximization, the interference mitigation [16,17], or the spectral efficiency improvement.

The approaches above concern the traditional cellular networks, where the users communicate via the eNB. Towards 5G networks and enabling the D2D communication capabilities of the users [18–20], respective RB allocation approaches can be found in [21–23]. In [21] the importance of RB allocation approaches in various and emerging network architectures is stressed via presenting different approaches in LTE and LTE-Advanced networks. In [22] the authors propose scheduling algorithms for the problem of RB allocation in D2D enabled networks taking into account each user's delay requirement. Here the eNB needs to determine the optimal assignment of resources to the users at time scales of milliseconds according to 3rd Generation Partnership Project (3GPP) specifications including devices that may communicate directly. In [23] the authors are proposing an efficient resource allocation scheme in clustered D2D enabled networks.

Focusing on the uplink RB allocation problem for the D2D enabled networks most of the related work is restricted to single cell environments [7–12], while very little attention has been paid so far to the respective multi-cell case. Specifically, in [24,25] the D2D RB selection problem is analysed as a game, in a simple scenario with one cellular user per cell. The players of the game are the BSs and they compete with each other in order to serve the D2D users with less harm. However, the existence of the Nash Equilibrium (NE) point of the game is not analytically proven, and it is only examined via numerical simulations.

The previous works focus on maximizing the total sum of rates in the network via game theoretic approaches and pricing for the resources. In [26], the authors address only the problem of RB allocation in a D2D underlay multicell network, via formulating a potential game with the objective of optimizing the total sum of rates in the network. In [27] the problem of total throughput maximization for the D2D users in a multicell D2D underlay cellular network is addressed under a Multiple Input Multiple Output (MIMO) setting. In [28] the problem of both RB and power allocation is addressed. Initially, the authors aim at minimizing the interference caused by the D2D links to the cellular links of the network, and subsequently a power control approach is applied in order to maximize total system throughput.

1.2. Paper Contributions

In this paper, we address the joint problem of optimal RB and power allocation in the uplink communication of multicell D2D enabled networks, where the optimization objective is the minimization of the total interference of the network. Due to the inherent complexity of the nature of the combined problem, a two step approach is followed. At the first step the RB allocation problem is solved considering that all users of the network transmit at fixed power. At the second step, taking the RB allocation vector of the first stage as an input, the total interference minimization problem is decomposed into multiple interference minimization problems per available RB.

Specifically, at the first step the problem is formulated as a non-cooperative game which is proven to be a bilateral symmetric interaction game. This type of games are exact potential games. Every player of the game is one of the potential receivers of the system. The receiver of the network can be either an eNB or a D2D receiver. The eNB is in charge of deciding the optimal RB assignment to the cellular users who belong to the respective cell. The D2D receiver decides the optimal RB for the respective D2D transmitter. Leveraging the highly desirable properties of potential games, we prove that at least one NE of the game exists. A best response algorithm is proposed which converges after a finite number of iterations to the identified NE point of the game.

It is noted that in our approach and methodology, in contrast to the majority of existing literature, we do not restrict a priori the number of D2D pairs that can reuse a RB, allocated to a cellular link. This is determined dynamically by the objective of interference minimization. Thus, a RB reserved by a cellular link can be potentially reused multiple times by different D2D links in order to increase spectrum efficiency.

During the second step of our approach each uplink power allocation problem per RB is formulated as a linear programming problem. This optimization problem is shown that is equivalent to the total power minimization problem for every available RB. For this reason, the well known Foschini-Miljanic algorithm [29], which solves the total power minimization problem [30], is properly adopted to solve our problem in a distributed way. This algorithm finds either a feasible solution or pushes the users who cannot satisfy their SINR requirement to the maximum transmission power level in order to achieve a SINR level as close as possible to their SINR requirement.

Finally, the efficiency of the overall proposed framework is evaluated through numerical simulations, especially with respect to significantly decreasing the total power consumption in the network, while increasing the network capacity in terms of number of users who are able to satisfy their SINR requirement.

2. System Model

We consider a multicell D2D underlay cellular network, where L neighbouring cells co-exist. The UEs and the eNBs are assumed to be equipped with a single omni-directional antenna. Each cell contains two types of users, cellular users (CUs) and D2D users (DUs), where their communication modes are assumed that have already been selected. The CUs communicate through the eNB of their serving cell, while the DUs communicate directly to each other in pairs, where the DUs of every D2D pair are adequately close to each other, thus forming a D2D link. The corresponding reference topology is shown in Figure 1.

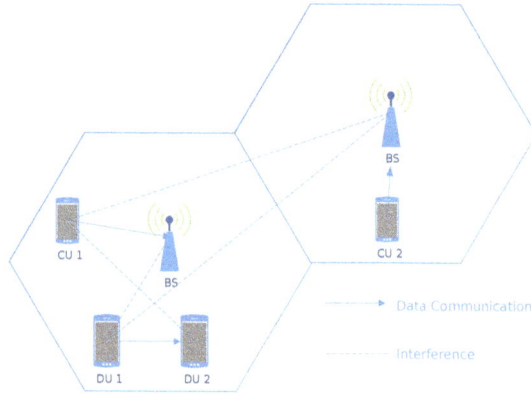

Figure 1. System Model.

We assume that each cell has access to a fixed number of K RBs. Every cellular user of each cell uses one of those RBs which are orthogonal to each other. This means that there is no interference between cellular users of the same cell. The number of cellular users at each cell requesting a RB can be variable from cell to cell. Considering that the set of cellular users of cell i will be requesting and assigned a non-overlapping set of RBs, each cellular user of the same cell will occupy a different RB. Apart from the cellular users, D2D users reside in the network and each of them has to decide which RB to choose in order to transmit. The D2D users reuse one of the available K RBs at each cell. Multiple D2D users may share the same RB with one cellular user in each cell. As mentioned before, RBs reuse takes place between cellular users from different cells and between cellular and D2D users from the same or different cell. In principle, a RB may be occupied simultaneously from only one cellular user per eNB, and from multiple D2D users.

In this multicell D2D underlay architecture, in order to treat cellular and D2D users in a homogeneous and holistic manner from RB allocation point of view, apart from the cellular cells i.e., eNBs, we consider the D2D receivers as virtual cells. Each D2D receiver can play the role of a (virtual) cell, where the serving users of this cell is exactly one i.e., the D2D transmitter of the respective D2D pair. From this point and on, when we refer to a cell, we may consider either the eNB of a cell or the D2D receiver of the respective D2D pair (that is we refer to either the physical L neighbouring cells or the considered virtual cells as defined above).

Let E be the set of all the cells (including the virtual ones) and $|E|$ be its cardinality that represents the number of the considered cells in the network. Let V_i be the set of users of cell i and let also K be the set of the available RBs at cell i which a user u can use. We define the SINR at the receiver of every user u of cell i, either cellular or D2D, using RB k, as follows:

$$\gamma_u^k = \frac{G_u^{i,k} \cdot P_u^k}{\sum\limits_{j \neq i} \sum\limits_{v \in V_j} G_v^{i,k} \cdot P_v^k + \sigma^2} \tag{1}$$

where $u \in V_i$, i.e., the set of users who belong to cell i, $G_u^{i,k}$ denotes the channel gain between user u of the cell i and the receiver-cell i at RB k, P_u^k is the transmission power of user u at RB k, V_j is the set of users who belong to cell j, $G_v^{i,k}$ is the channel gain between user v of the cell j and the receiver of cell i at RB k, P_v^k is the transmission power of user v at RB k and σ^2 refers to the thermal background noise power at the receiver of each link.

3. Problem Formulation

In the following, we formulate the problem of interference minimization in the joint space of RBs and uplink transmission power.

$$\underset{S,P}{\text{minimize}} \sum_{i=1}^{E} I_i \tag{2}$$

$$\text{subject to } \sum_{k=1}^{K} s_i(u,k) = 1 \left.\begin{array}{l}\\ \\ \\\end{array}\right\} \tag{2a}$$

$$0 \leq P_u \leq P_u^{max} \quad \forall u \in V_i, \forall i \in E \tag{2b}$$

$$\gamma_u \geq \gamma_u^{tar} \tag{2c}$$

$$\sum_{u \in V_i} s_i(u,k) = 1, \quad \forall k \in K, \forall i \in E \tag{2d}$$

where

$$I_i = \sum_{u \in V_i} \sum_{k=1}^{K} s_i(u,k) \sum_{j \neq i} \sum_{v \in V_j} s_j(v,k) \cdot P_v \cdot G_v^{i,k} + \sigma^2 \tag{3}$$

is the total sensed interference from all users at cell i, caused by any other user of the other cells of the system who share common RBs with the users of cell i and

$$s_i(u,k) = \begin{cases} 1 & \text{if user } u \text{ of cell } i \text{ uses RB } k \\ 0 & \text{otherwise} \end{cases} \tag{4}$$

User u belongs to cell i and user v belongs to cell j. P_u denotes the transmission power of user u of the cell i at the chosen RB from user u. Since every user is assumed to use exactly one RB we can omit the exponent k from P_u^k without disambiguation. P_u^{max} is the maximum transmission power of user u. γ_u is the SINR at the receiver of user u. Similarly the exponent k is omitted from γ_u^k. γ_u^{tar} is a positive real number denoting the SINR requirement of user u.

$S = \{s_i\}_{i=1,\dots,|E|}$ and $P = \{P_u\}_{u \in V_i, i=1,\dots,|E|}$, where $|E|$ is the number of the cells in the network.

The first Constraint (2a) concerns the fact that each user of each cell (either physical or virtual) chooses exactly one RB. Constraints (2b) and (2c) concern the acceptable range of values for the transmission power and SINR levels. The final Constraint (2d) is related to the orthogonality requirement between users of the same cell. It is stressed here that based on the consideration of the additional virtual cells, a D2D user belongs to the corresponding cell (which essentially is a single user cell) and not to the physical cell that is placed within. In a nutshell, in order to solve the aforementioned optimization problem we search for feasible combinations of RB and transmission power assignments which minimize the total interference in the network.

The above problem is a mixed integer programming problem since some of the decision variables are integer numbers and the rest of them i.e. transmission powers, continuous variables. This problem is a combinatorial optimization problem and its complexity is known to be NP-hard [31]. For this reason, we aim at efficient approximation algorithms to solve the problem. The proposed approach is to solve the problem in two stages. In the first stage, we consider that all users have chosen a fixed transmission power and we search for the optimal RB assignment. We propose a game theoretic approach which solves the problem iteratively. At each iteration every player of the game makes a decision regarding its RB which maximizes its utility function while simultaneously minimizes the total interference in the network. After a finite number of iterations the algorithm converges to the desirable NE point of the game. At the second stage, the problem of total interference minimization can be decomposed to multiple interference minimization problems per RB. Certain number of users share a RB and create interference one to another. Our goal is to determine users' optimal transmission power allocation towards minimizing the total interference per RB while satisfying the minimum SINR requirements of the respective users.

4. RB Allocation

4.1. RB Problem Formulation and Solution

In this subsection we describe the first stage of our proposed approach which concerns the RB allocation. It is assumed that every user of the cell (either cellular or D2D) has a fixed transmission power value. Thus the problem, expressed in Equation (2), may be simplified and take the following form

$$\underset{S}{\text{minimize}} \sum_{i=1}^{E} I_i \tag{5}$$

$$\text{subject to} \sum_{k=1}^{K} s_i(u,k) = 1, \forall u \in V_i, \forall i \in E \tag{5a}$$

$$\sum_{u \in V_i} s_i(u,k) = 1, \forall k \in K, \forall i \in E \tag{5b}$$

$$\gamma_u \geq \gamma_u^{tar}, \forall u \in V_i, \forall i \in E \tag{5c}$$

where I_i is defined as in Equation (3) and P_v has been assumed a fixed value for every $v \in V_j$ and $j \in E$.

The above problem is a combinatorial optimization problem since the decision variables of the problem i.e., the assignment matrices of each cell are integer. In the literature, efficient algorithms for the exact solution of this type of problems do not exist. For this reason, we search for approximation algorithms which provide suboptimal solutions in a time efficient manner.

For this problem, our idea is to model it, as an exact potential game. The players of this game will be the cells—either physical or virtual (D2D)—of the system and the strategies will be the assignment matrices of the cells. Formally, we define the game

$$G = \{N, S, U\} \tag{6}$$

where

- N is the set of the players of the game G, $N = E$, i.e., the set of the receivers in the system—either the eNBs or the D2D receivers,
- S denotes the set of the strategies of the players, $S = \{s_i\}_{i=1,\dots,|N|}$ i.e., the set of the assignment matrices of each cell-receiver, s_i, defined in Equation (4) and
- U denotes the set of the utility functions of the players of the game G.

The utility function U_i of the player i is defined as follows:

$$U_i(s_i, s_{-i}) = -I_i(s_i, s_{-i}) - I_i^c(s_i, s_{-i}) \tag{7}$$

where $I_i(s_i, s_{-i})$ is defined in Equation (3) and repeated here for clarity reasons

$$I_i = \sum_{u \in V_i} \sum_{k=1}^{K} s_i(u,k) \sum_{j \neq i} \sum_{v \in V_j} s_j(v,k) \cdot P_v \cdot G_v^{i,k} + \sigma^2$$

$I_i^c(s_i, s_{-i})$ is defined as follows

$$I_i^c(s_i, s_{-i}) = \sum_{u \in V_i} \sum_{k=1}^{K} s_i(u,k) \cdot P_u \sum_{j \neq i} \sum_{v \in V_j} s_j(v,k) \cdot G_u^{j,k} + \sigma^2 \tag{8}$$

$I_i^c(s_i, s_{-i})$ denotes the total interference caused by all users of cell i to all other cells using at least one common RB with cell i.

Theorem 1. *Game G is a bilateral symmetric interaction game [32].*

Proof of Theorem 1. A bilateral symmetric interaction game has the following form $B = \{N, S, U\}$, where

$$U_i(s_i, s_{-i}) = \sum_{j \neq i} w_{i,j}(s_i, s_j), \forall i \in N \qquad (9)$$

and

$$w_{i,j}(s_i, s_j) = w_{j,i}(s_j, s_i), \forall i \neq j \qquad (10)$$

where $w_{i,j} : S_i \times S_j \to \mathbb{R}$.

The utility function of the player i of the game G, defined in Equation (7), can be mapped to the utility function defined in Equation (9), if $w_{i,j}(s_i, s_j) = -\sum_{u \in V_i}\sum_{k=1}^{K} s_i(u,k) \sum_{v \in V_j} s_j(v,k)(P_v G_v^{i,k} + P_u G_u^{j,k}) + \sigma^2$. It is observed that the bilateral symmetric property of Equation (10) also holds. $\square$

Bilateral symmetric interaction games exhibit certain desirable properties since they are exact potential games. The potential function of the game G according to [32] is the following:

$$Pot(s_i, s_{-i}) = \sum_{i \in N}\sum_{j \in N, j < i} w_{i,j}(s_i, s_j) = -\sum_{i \in N}\sum_{j \in N, j < i}\sum_{u \in V_i}\sum_{k=1}^{K} s_i(u,k) \cdot \sum_{v \in V_j} s_j(v,k)(P_v G_v^{i,k} + P_u G_u^{j,k}) \quad (11)$$

We observe that there is a symmetry between the interactions of one player i and another player j. This holds, because the total interference sensed and caused by the users of cell i to the cell j is equal to the respective interference from the users of cell j to the cell i. We characterize as a non overlapping set of users, the users who are allocated orthogonal RBs. The interference sensed by a non overlapping set of users of the cell i, due to the cell j, is the total interference that a non overlapping set of users of the cell j create to the cell i. The inverse is also true.

Since the game G is an exact potential game, this means that this game has at least one NE point and this NE point is a local optimum of the potential function defined in Equation (11). The potential function, defined in Equation (11), can be easily proven (change of variables i and j for the second term) that it coincides with minus the total interference sensed by every cell in the network. Thus, every NE point of the game G is a local minimum of the total interference.

4.2. RB Allocation Iterative Algorithm

In this subsection, we will present a best response algorithm which is proven to converge after a finite number of iterations to one of the NEs of the game G. We consider that initially all players of the game, i.e., the cells, have chosen an initial RB assignment for all users who reside inside them. Below the respective algorithm is presented.

In Algorithm 1, every player of the game, i.e., each cell acts in a sequential order. This assures that Algorithm 1 will converge to the NE of the game i.e., minimum of the total interference after a finite number of iterations. It chooses the optimal RB assignment to its users, based on the equation of Line 4 in the Algorithm 1. All other users are considered that they have already chosen a RB assignment to their users and then the current cell according to Line 4 chooses the new optimal RB assignment for its users. This step is repeated until all players of the game conclude to a common RB assignment for all users of the network.

Algorithm 1: RB allocation

1: Counter $r = 0$, Converged = false, ite(i) = $0 \: \forall i \in N$
2: **for all** cells **do**

3:   **if** *Converged* $== false$ **then**

4:     $s_i(ite + 1) = \arg \max_{s_i} U_i(s_i(ite(i)), s_{-i}(ite(-i)))$
5:     **if** $s_i(ite(i) + 1) == s_i(ite(i))$ **then**

6:       $r = r + 1$
7:     **else**

8:       $r = 0$
9:     **end if**
10:     $ite(i) = ite(i) + 1$
11:     **if** $r == |N| - 1$ **then**

12:       *Converged* = *true*
13:     **end if**
14:   **end if**
15: **end for**

More precisely, from Equation (7) and considering that all other cells, apart from cell i, have chosen RB assignment the decision problem of cell i takes the following form:

$$\underset{S}{\text{maximize}} \: U_i(s_i, s_{-i}) = - \sum_{u \in V_i} \sum_{k=1}^{K} s_i(u, k) b_i(u, k) \tag{12}$$

$$\text{subject to} \sum_{k=1}^{K} s_i(u, k) = 1, \forall u \in V_i \tag{12a}$$

$$\sum_{u \in V_i} s_i(u, k) = 1, \forall k \in K \tag{12b}$$

where

$$b_i(u, k) = \sum_{j \neq i} \sum_{v \in V_j} s_j(v, k) (P_v G_v^{i,k} + P_u G_u^{j,k}) \tag{13}$$

and $u \in V_i$.

Problem (12) is an integer linear programming problem and even the relaxed version of it always has an integer optimal value since the constraint matrix is totally unimodular. This problem can be solved via the well known Hungarian algorithm [33] in polynomial time with respect to the number of the available RBs at each cell. Leveraging the potential game properties, Algorithm 1 converges after a finite number of iterations. Moreover, at each step, Algorithm 1 improves not only the utility function of each cell but also the potential function of the game i.e., the total interference. After a finite number of iterations, a RB assignment is selected which minimizes the total interference of the network.

In order for Algorithm 1 to be practically implementable, there is a need for a communication control channel between the different cells i.e., the eNBs and the D2D cells. This control channel will carry the necessary information in order for the cells to execute Algorithm 1. For the case of eNBs, this can be done using the X2 interface [34] which has been already designed to implement this type of communication between the eNBs in LTE networks. For the D2D cells it can be assumed that there is a control channel between the D2D receivers and the respective serving eNB they reside in [19]. The complexity of Algorithm 1 belongs to the class of Polynomial local search (PLS) problems [35]. At each step, each eNB runs the Hungarian algorithm in order to update its RB assignment. The complexity of the Hungarian algorithm is $O(K^3)$ with respect to the number of the available RBs. For the D2D cells the complexity of the update step is $O(K)$ since the number of served users is only one i.e., the respective D2D transmitter. The convergence of the above algorithm is guaranteed but not necessarily in polynomial time. However, at every step of the Algorithm 1 the total interference is being reduced. As it will be seen in Section 6.1.1 the required number of iterations for convergence was quite low with respect to the number of players involved. By applying Algorithm 1, we will obtain a RB assignment for all cells of the network which minimizes the total interference in the network.

5. Power Control

In this section, we present the second stage of our approach for the power control problem in multicell D2D underlay network architectures, considering that the RBs have already been assigned to the cellular and the D2D users according to the output of the first stage of our approach. Thus, capitalizing on the outcome of the RB allocation approach, Problem (2) takes the following form:

$$\underset{P}{\text{minimize}} \sum_{i=1}^{E} I_i \tag{14}$$

$$\left. \begin{array}{l} \text{subject to } 0 \le P_u \le P_u^{\text{max}} \\ \gamma_u \ge \gamma_u^{tar} \end{array} \right\} \forall u \in V_i, \forall i \in E \tag{14a} \tag{14b}$$

All cells have already chosen their RB assignment matrices and thus Problem (14) can be decomposed to multiple optimization problems per available RB since we assume that there is no interference between different RBs due to their orthogonality. Problem (14) is equivalent to multiple interference minimization problems per RB k. Then each problem for every RB k can be written as:

$$\underset{P}{\text{minimize}} \sum_{i \in R_k} \sum_{\substack{j \in R_k, \\ j \ne i}} G_v^{i,k} P_v + \sigma^2 \tag{15}$$

$$\left. \begin{array}{l} \text{subject to } 0 \le P_v \le P_v^{\text{max}} \\ \gamma_v \ge \gamma_v^{tar} \end{array} \right\} \forall j \in R_k, \forall v \in V_j \tag{15a} \tag{15b}$$

R_k is the set of cells who occupy RB k and v the user of cell j who occupies RB k. There is no other user who uses the same RB at the same cell since we assume orthogonality of the available RBs. Let P_v be the transmission power of the the user v of cell j who uses RB k.

The objective is to minimize total interference at every RB k under the constraints of feasible transmission power and SINR requirement satisfaction of the users. Problem (15) is a typical linear programming problem. Thus, if this problem has a feasible solution this lies at the point of intersection of the Constraints (15a) and (15b). The gradient of the objective function shows the direction of the optimal change of the objective Function with respect to the independent variables of the problem i.e., transmission powers of the users sharing RB k. The gradient of the objective Function (15) is positive with respect to the transmission power of every user of each cell who uses common RB k. Thus, the minimum power which satisfies all Conditions (15a) and (15b) is the optimal feasible solution which minimizes the objective function of Problem (15).

The gradient of the objective function of Problem (15) has the same sign to the total power minimization problem under the same SINR and transmission power constraints. For the problem of total power minimization, as it is already known from the literature [29], there is a distributed algorithm which converges to the optimal solution, if this solution is feasible i.e., satisfies Constraints (15a) and (15b). Thus, by applying this algorithm to Problem (15), we can find the required optimal solution.

Let P_{init} be the initial transmission power vector of all users of the network. This is a fixed transmission power vector which could be the maximum transmission power value for each user of the network. At Line 4 of Algorithm 2, each user updates its current power in a parallel fashion. After the execution of this step, a coordination mechanism is needed in order to check the power convergence of each user.

Algorithm 2: Distributed Power Control per RB k

1: Counter $t = 0$, *Converged* $= false$, $P(0) = P_{init}$
2: **if** *Converged* $== false$ **then**

3:     **for all** $i \in R_k$ **do**

4:         $P_u(t+1) = \min \left\{ \frac{\gamma_u^{tar}}{\gamma_u(t) P_u(t)}, P_u^{max} \right\}, u \in V_i, s_i(u, k) = 1$
5:     **end for**
6:     **if** $P_{u_i}(t+1) == P_{u_i}(t) \ \forall i \in R_k$ **then**

7:         *Converged* $= true$
8:     **end if**
9:     $t = t + 1$
10: **end if**

---

6. Performance Evaluation

In this section, we provide a series of numerical results evaluating the operational features and the performance of the proposed approach. Initially, in Section 6.1, we focus on the operation performance achievements of our proposed methodology, in terms of interference, uplink transmission power and achievable data rate. The achievable performance and contribution of the proposed RB allocation process (i.e., first step) is examined and evaluated in Section 6.1.1, while subsequently in Section 6.1.2 the additional benefits obtained by the second stage of our approach (i.e., power control) are studied and its fast convergence to the final optimal power values is shown. Then, in Section 6.2 we provide a comparative evaluation of our proposed approach against another existing approach in the recent literature that targets rate maximization [26], through the efficient resource allocation in a multi-cell network supporting D2D communication. The aforementioned approaches are thoroughly compared in order to illustrate the obtained gain of our proposed approach in terms of uplink transmission power required to meet the users' SINR requirements, number of users that can be supported by each approach (i.e., users that their SINR meets or exceeds the pre-specified threshold), rate and interference. It should be noted that unless otherwise is explicitly mentioned, the majority of the numerical results is the statistical average outcome of several random topologies with respect to the distribution of the cellular and D2D users within each cell.

6.1. Proposed Approach Properties and Operation

6.1.1. RB Allocation

In this subsection, the effectiveness and efficiency of the first step of our approach i.e., Algorithm 1 is shown. The simulation setup and the corresponding parameters considered are shown in Table 1.

Table 1. Simulation Setup.

Parameters	Values
Number of evolved NodeBs (eNBs)	3
Number of available Resource Blocks (RBs) per eNB	10
Number of Device to Device (D2D) cells per eNB	16
Number of cellular users per eNB	8
Cellular Cell Radius	300 m
D2D Cell Radius	30 m
Channel Gains Loss Exponent	3

We consider an initial random RB assignment to all users of all cells (both cellular and D2D users) of the network, where each user is assumed to use exactly one RB. Cellular users of the same cell use different RBs in order to satisfy the orthogonality requirement for every cell. The channel gains between a user u of cell i and the receiver at cell j is computed as $G_{u_i}^j = 1 / (d_{u_i}^j)^3$. In Figure 2, the specific topology under consideration in this experiment along with the initial RB assignment is presented. The corresponding numbers represent the RB assigned to the respective users, therefore this figure practically shows a graphical representation of the reuse of each RB. In Figures 2 and 3, the numbers without the special character ′ refer to cellular users while the numbers followed by ′ refer to D2D users.

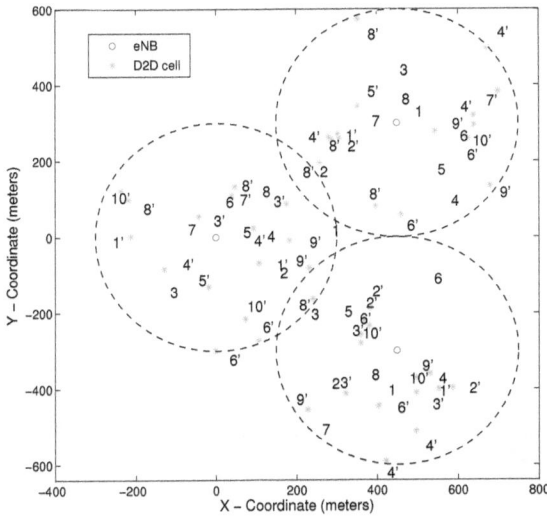

Figure 2. Topology of the initial Resource Block (RB) allocation.

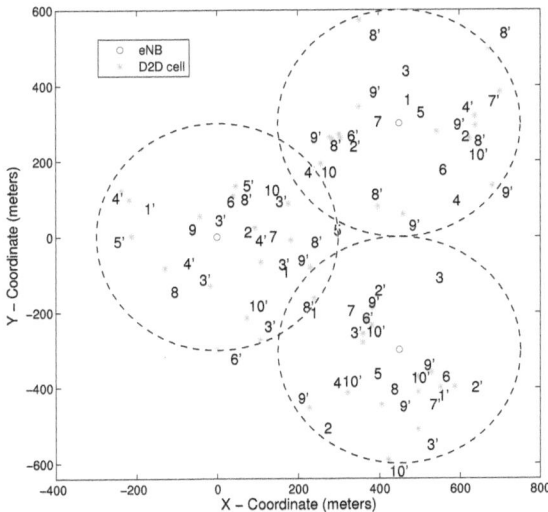

Figure 3. Topology of the final RB allocation.

Similarly, in Figure 3, the final RB allocation as the outcome of the application of the first step of our approach (i.e., Algorithm 1) to all users of the network is shown. It is clearly demonstrated that the RB reuse efficiency has been significantly improved. Specifically, considering as metric of such efficiency the achieved reuse distance of a common RB—denoting the distance between a transmitter and a different D2D receiver who share the same RB—it is observed that the reuse distance of all RBs increases drastically. Precisely, the average minimum reuse distance of every RB, before the application of Algorithm 1, was approximately 15 m and after the application of Algorithm 1 it increases to approximately 35 m, corresponding to an improvement of 133%.

In Figure 4, the evolution of the total interference in the network is demonstrated with respect to the number of iterations needed for Algorithm 1 to converge. It is shown that Algorithm 1 always reduces the total interference after each step of its execution. This also verifies experimentally the respective theoretical property of potential games [32], as the one considered in this paper. The required number of iterations for convergence was quite low with respect to the total number of players of the game. Precisely, the required number was about 3–4 times the total number of players involved in the game. The total interference reduction compared to the corresponding value based on the initial random allocation is approximately 80%. It is noted that we have also verified experimentally that Algorithm 1 converges to a stable outcome under any initial RB allocation, while the range of the optimal values of the total interference was within 5% difference of the observed maximum value.

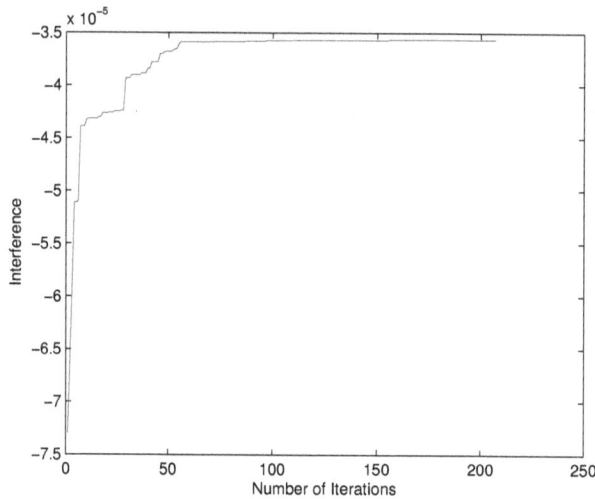

Figure 4. Total Interference Evolution.

Subsequently, in Figure 5, we show the average interference reduction for every user of the network after the application of Algorithm 1. For clarity reasons, it is mentioned that users 1–24 in all figures represent cellular users, while the users 25 and beyond represent D2D users.

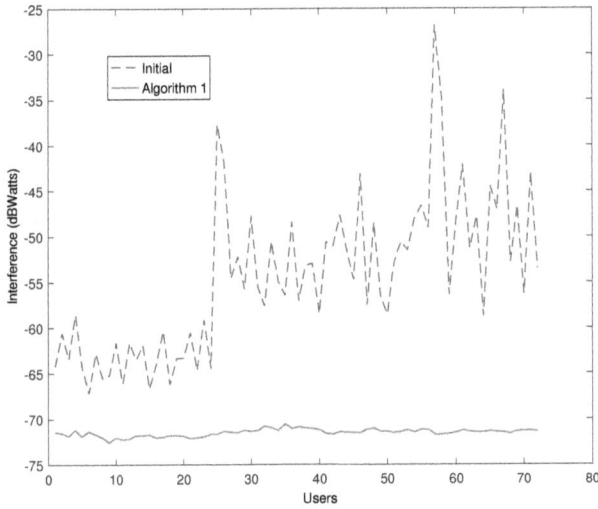

Figure 5. Interference before and after the application of Algorithm 1.

Based on these results, we observe that Algorithm 1 achieves a significant reduction to the interference, sensed at every user of the network. Precisely, the mean interference reduction is on average 93%. This in turn can be translated to significant improvements in the achievable transmission rate as well, as it is observed in Figure 6, where the achieved transmission rate per used bandwidth unit (i.e., spectral efficiency) is shown for every user of the network. The mean rate increase with respect to the initial achievable transmission rate is calculated at approximately 34% .

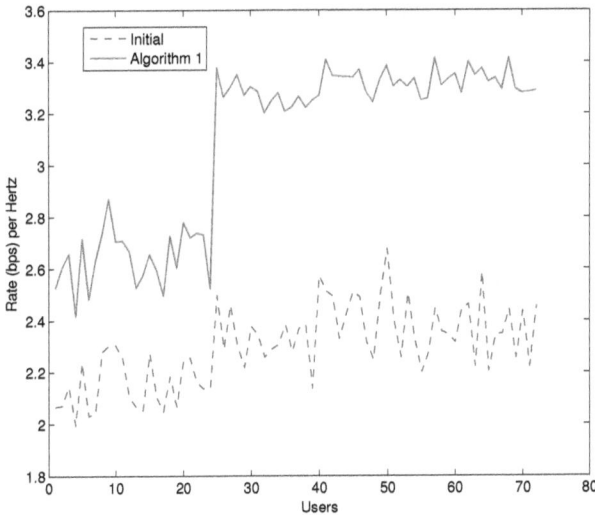

Figure 6. Rate per Hertz before and after the application of Algorithm 1.

6.1.2. Power Allocation

In this subsection, the second step of our approach i.e. the power allocation Algorithm 2 is evaluated. After the application of Algorithm 1, a RB selection for all users of the network is obtained.

Then as explained before, we apply Algorithm 2 for every co-sharing RB. The simulation setup for this section of our experimentation is the same as presented in Table 1. It is noted that in principle users may have different SINR requirements depending on the type of service they request. For demonstration only purposes in the set of results presented here, we have assumed a common SINR requirement for all users of the network equal to 10. Algorithm 2 instructs users who share a common RB to act in a parallel fashion. The convergence of Algorithm 2 takes place after a quite low number of iterations. All users start from an initial power allocation which is their maximum transmission power. This is considered to be 2 watts for the cellular users and 0.1 watts for the D2D users. During the convergence process of Algorithm 2, each user tries to find the minimum necessary transmission power in order to satisfy his SINR requirement. In case this is not found, it transmits at his maximum power.

In Figure 7, the transmit power convergence is shown for a randomly selected RB, that is RB 6. It is shown that all users of the network who share RB 6 converge to their desired transmission power values. All users of this RB either meet or exceed their SINR requirement. The transmit power radically reduces for all users, and as a result the total power consumption decreases by 57% with respect to the initial power values. Moreover the average number of users, throughout the whole network, who satisfy their SINR requirement increases from 29.15 to 55.35 after the application of Algorithm 1 corresponding to an increase of approximately 90%, and finally to 61.19 users after the combined application of Algorithms 1 and 2, raising this improvement to 110% in total.

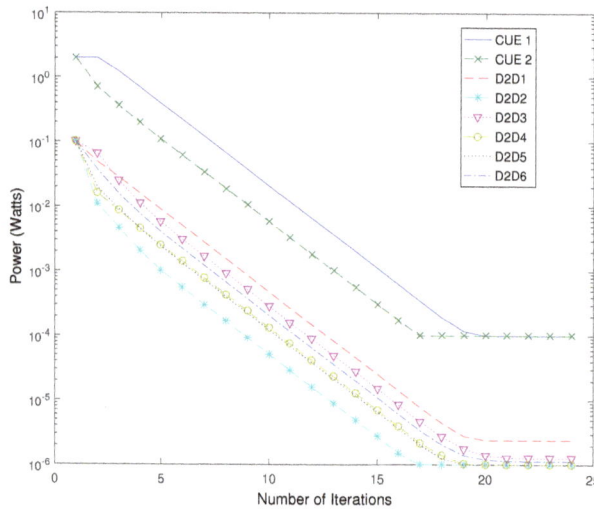

Figure 7. Transmit Power Convergence for a random RB.

6.2. Comparative Results

In this subsection our proposed approach is compared against a state of the art approach in the recent literature [26]. The latter concerns the RB allocation in a multicell D2D underlay network, with the objective of optimizing the total sum of transmission rates of all users of the network, through an approximate better response algorithm under a potential game formulation. In the following, we refer to this approach as Rate Max approach.

The comparative process and evaluation is performed in a gradual manner in order to clearly demonstrate and quantify the benefits obtained by the application of each step of our approach, while maintaining fairness in the comparisons. Our approach aims at total interference minimization while the approach of [26] aims at throughput maximization in the RB allocations space. The first step of our approach i.e., Algorithm 1 achieves on average better results with respect to the interference at the receiver of each user compared to [26]. This gives the users the opportunity as it will be shown next, by applying Algorithm 2 to transmit at lower levels on average while a larger number of them satisfies their SINR requirements. Specifically, the following four scenarios are considered and compared:

- the application of only Algorithm 1 i.e., our proposed algorithm for the RB allocation phase of the problem
- the application of approach [26] for the RB allocation phase of the problem
- the combined application of Algorithms 1 and 2
- the combined application of approach [26] and Algorithm 2 of our approach (for power control).

In this subsection, the above approaches are tested and evaluated under the simulation setup used in previous Section 6.1.2, that is the one presented in Table 1 with the SINR requirement equal to 10. As mentioned before, all the numerical results presented in the following refer to statistical average values, as they are obtained by several repetitions of the simulation, with the locations of the cellular and D2D users varying uniformly and randomly in the network topology. The combined application concerns the sequential application of a RB allocation approach and then the application of our proposed power control approach (i.e., Algorithm 2).

In Figure 8, the first step of our approach i.e. Algorithm 1 is compared directly to approach [26] in terms of observed interference. As clearly indicated by the corresponding results Algorithm 1 achieves a considerable interference reduction compared to [26]. Specifically, the average reduction of our approach with respect to [26] is calculated as approximately 50% per user. It should be clarified that this value refers to the net value of interference, while the vertical axis of Figure 8 is expressed in decibel Watts (dBWatts). It can be seen that the largest benefit concerns the cellular users, where the respective reduction is almost 80%. In Figure 9, the respective transmission rates per Hertz for each user are shown. It can be seen that although Algorithm 1 targets at interference minimization, under fixed transmission power obtains comparable to [26] performance with respect to the achievable transmission rate.

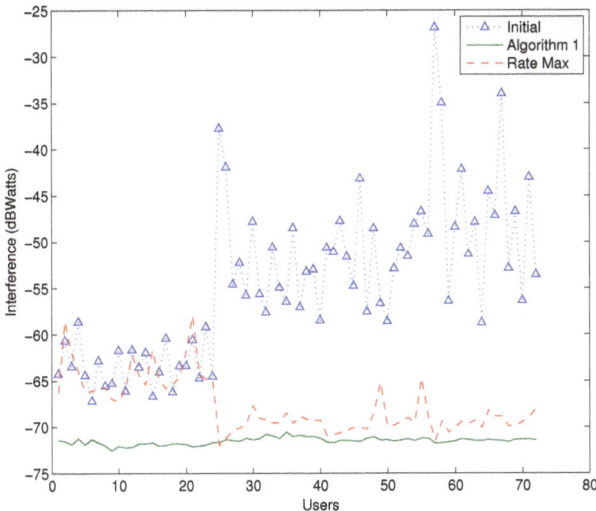

Figure 8. Comparison of interference (initial, Algorithm 1, Rate Max [26]).

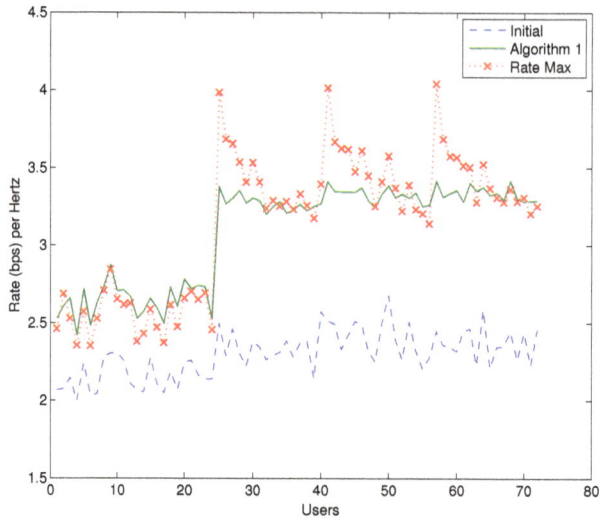

Figure 9. Comparison of Rate per Hertz (initial, Algorithm 1, Rate Max [26]).

By applying Algorithm 2 after the execution of Algorithm 1 and approach [26] respectively, we compare and analyse the averaging results with respect to the transmission power and the number of users satisfying their SINR requirements. In Figure 10, it is demonstrated that the required transmission power under our proposed approach reduces radically, i.e., by 70% on average with respect to the initial transmission power values and by 30% with respect to the combined approach (Rate Max + Algorithm 2). Lastly, the number of users in the system who are able to satisfy their SINR requirement also increases following both approaches, with our approach outperforming by 12% the (Rate Max + Algorithm 2) (Figure 11).

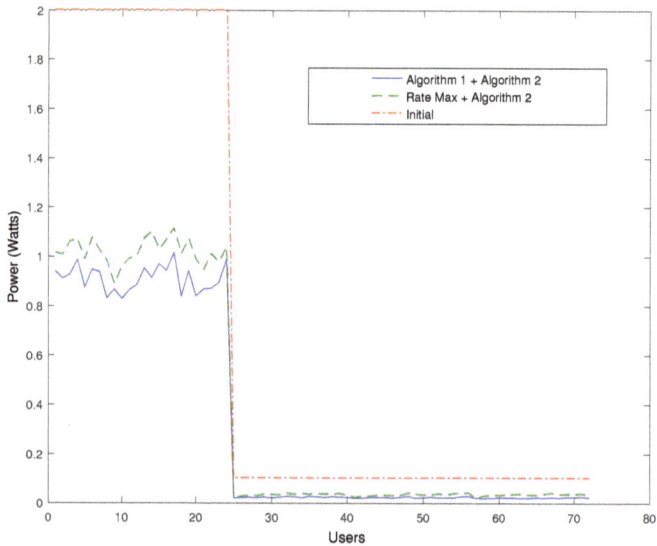

Figure 10. Comparison of the transmission power for the combined approaches.

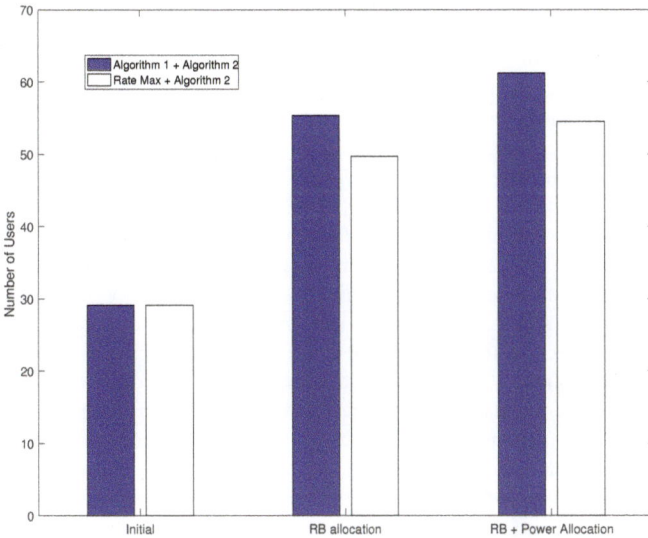

Figure 11. Average number of satisfied users (in terms of Signal to Interference and Noise Ratio (SINR) requirement) after each stage of the Resource Block (RB) and power allocation procedure.

Similarly in Figure 12, the total number of users who satisfy their SINR requirements as a function of increasing number of D2D users per eNB is presented. It is clearly observed that our approach sustains its significant capacity improvement in terms of number of users that can be supported by the network resources even both for low and high density networks (i.e., increasing number of D2D users per eNB).

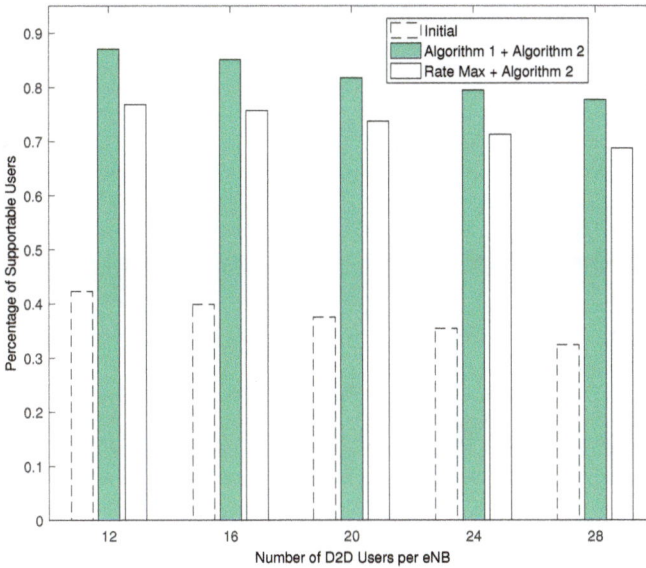

Figure 12. Average number of satisfied users (in terms of SINR requirement) for the combined approaches, for increasing number of Device to Device (D2D) users per evolved NodeB (eNB).

7. Conclusions

In this paper, the problem of interference mitigation in a multi-cell D2D underlay cellular network is addressed. To deal with the complexity introduced by the multiple cell environment and the existence of D2D users, a two step approach has been proposed and adopted. As the first step, the RB allocation problem is solved by formulating it as a non-cooperative game which is proven to be a bilateral symmetric interaction game. We show that at least one NE of the game exists and a best response algorithm is introduced which converges after a finite number of iterations to the identified NE point of the game. Subsequently, at the second step the total interference minimization problem is decomposed into multiple interference minimization problems per available RB, and power control is applied by each user in a distributed manner in order to identify the lowest required transmission power to satisfy its SINR, if possible. Extensive numerical results are presented that verify several properties and operational characteristics of our proposed approach. Specifically it is demonstrated that our proposed approach manages to reduce significantly the total interference in the network as well as the interference at the receiver of every user, either cellular or D2D. Furthermore, by comparing our approach to another existing approach in the recent literature that targets rate maximization, it is shown that though our approach targets interference minimization, it still obtains comparable performance in terms of total throughput, and as a consequence significant spectral efficiency improvement is finally achieved. This enhanced performance translates directly to system capacity increase in terms of number of users that can be supported by the network resources and transmission power reduction.

Author Contributions: All authors contributed extensively to the work presented in this paper. Symeon Papavassiliou had the original idea on which we have based our current work, supported the design and orchestration of the performance evaluation work, and had the overall coordination in the writing of the article. Georgios Katsinis led the work in the optimization problem formulation and solution, defined the proposed two-step algorithm, implemented the code for the conduction of the simulations and had key leading role in the article writing. Eirini Eleni Tsiropoulou, helped mainly in the formulation and solution of the optimization problem through a game theoretic approach, while contributed to the discussions and analysis of the comparative evaluation results and the overall editing of the paper.

Conflicts of Interest: The authors declare no conflict of interest.

References

1. Hasan, S.F. 5G communication technology. In *Emerging Trends in Communication Networks*; Springer: Berlin/Heidelberg, Germany, 2014; pp. 59–69.
2. 3rd Generation Partnership Project (3GPP). Study on LTE Device to Device Proximity Services; Radio Aspects. TR 36.843, 2014 Available online: http://www.tech-invite.com/3m36/tinv-3gpp-36-843.html (accessed on 4 August 2017).
3. Asadi, A.; Wang, Q.; Mancuso, V. A survey on device-to-device communication in cellular networks. *IEEE Commun. Surv. Tutor.* **2014**, *16*, 1801–1819.
4. Datsika, E.; Antonopoulos, A.; Zorba, N.; Verikoukis, C. Cross-Network Performance Analysis of Network Coding Aided Cooperative Outband D2D Communications. *IEEE Trans. Wirel. Commun.* **2017**, *16*, 3176–3188.
5. Antonopoulos, A.; Kartsakli, E.; Verikoukis, C. Game theoretic D2D content dissemination in 4G cellular networks. *IEEE Commun. Mag.* **2014**, *52*, 125–132.
6. Access, E.U.T.R. *Physical Channels and Modulation (3GPP TS 36.211 Version 12.4. 0 Release 12)*; ETSI: Sophia-Antipolis, France, 2014; Volume 12.
7. Katsinis, G.; Tsiropoulou, E.E.; Papavassiliou, S. Joint Resource Block and Power Allocation for Interference Management in Device to Device Underlay Cellular Networks: A Game Theoretic Approach. *Mob. Netw. Appl.* **2016**, *22*, 539–551.
8. Feng, D.; Lu, L.; Yuan-Wu, Y.; Li, G.; Li, S.; Feng, G. Device-to-device communications in cellular networks. *IEEE Commun. Mag.* **2014**, *52*, 49–55.
9. Phunchongharn, P.; Hossain, E.; Kim, D.I. Resource allocation for device-to-device communications underlaying LTE-advanced networks. *IEEE Wirel. Commun.* **2013**, *20*, 91–100.

10. Yu, C.H.; Doppler, K.; Ribeiro, C.B.; Tirkkonen, O. Resource sharing optimization for device-to-device communication underlaying cellular networks. *IEEE Trans. Wirel. Commun.* **2011**, *10*, 2752–2763.

11. Shah, S.T.; Gu, J.; Hasan, S.F.; Chung, M.Y. SC-FDMA-based resource allocation and power control scheme for D2D communication using LTE-A uplink resource. *EURASIP J. Wirel. Commun. Netw.* **2015**, *2015*, 137.

12. Da Silva, J.M.B.; Fodor, G. A binary power control scheme for D2D communications. *IEEE Wirel. Commun. Lett.* **2015**, *4*, 669–672.

13. Yaacoub, E.; Dawy, Z. A survey on uplink resource allocation in OFDMA wireless networks. *IEEE Commun. Surv. Tutor.* **2012**, *14*, 322–337.

14. Kwan, R.; Leung, C. A survey of scheduling and interference mitigation in LTE. *J. Electr. Comput. Eng.* **2010**, *2010*, 1, doi:10.1155/2010/273486.

15. Wang, X.; Yang, Y.; Sheng, J. Energy Efficient Power Allocation for the Uplink of Distributed Massive MIMO Systems. *Futur. Internet* **2017**, *9*, 21.

16. Kosta, C.; Hunt, B.; Quddus, A.U.; Tafazolli, R. On interference avoidance through inter-cell interference coordination (ICIC) based on OFDMA mobile systems. *IEEE Commun. Surv. Tutor.* **2013**, *15*, 973–995.

17. Boudreau, G.; Panicker, J.; Guo, N.; Chang, R.; Wang, N.; Vrzic, S. Interference coordination and cancellation for 4G networks. *IEEE Commun. Mag.* **2009**, *47*, doi:10.1109/MCOM.2009.4907410.

18. Tehrani, M.N.; Uysal, M.; Yanikomeroglu, H. Device-to-device communication in 5G cellular networks: Challenges, solutions, and future directions. *IEEE Commun. Mag.* **2014**, *52*, 86–92.

19. Fodor, G.; Dahlman, E.; Mildh, G.; Parkvall, S.; Reider, N.; Miklós, G.; Turányi, Z. Design aspects of network assisted device-to-device communications. *IEEE Commun. Mag.* **2012**, *50*, doi:10.1109/MCOM.2012.6163598.

20. Militano, L.; Orsino, A.; Araniti, G.; Iera, A. NB-IoT for D2D-Enhanced Content Uploading with Social Trustworthiness in 5G Systems. *Futur. Internet* **2017**, *9*, 31.

21. Abu-Ali, N.; Taha, A.E.M.; Salah, M.; Hassanein, H. Uplink scheduling in LTE and LTE-advanced: Tutorial, survey and evaluation framework. *IEEE Commun. Surv. Tutor.* **2014**, *16*, 1239–1265.

22. Lioumpas, A.S.; Alexiou, A. Uplink scheduling for Machine-to-Machine communications in LTE-based cellular systems. In Proceedings of the 2011 IEEE GLOBECOM Workshops (GC Wkshps), Houston, TX, USA, 5–9 December 2011; pp. 353–357.

23. Yang, H.; Seet, B.C.; Hasan, S.F.; Chong, P.H.J.; Chung, M.Y. Radio Resource Allocation for D2D-Enabled Massive Machine Communication in the 5G Era. In Proceedings of the 2016 IEEE 14th International Conference on Dependable, Autonomic and Secure Computing, 14th International Conference on Pervasive Intelligence and Computing, 2nd International Conference on Big Data Intelligence and Computing and Cyber Science and Technology Congress (DASC/PiCom/DataCom/CyberSciTech), Auckland, New Zealand, 8–12 August 2016; pp. 55–60.

24. Huang, J.; Yin, Y.; Sun, Y.; Zhao, Y.; Xing, C.C.; Duan, Q. Game theoretic resource allocation for multicell D2D communications with incomplete information. In Proceedings of the 2015 IEEE International Conference on Communications (ICC), London, UK, 8–12 June 2015; pp. 3039–3044.

25. Huang, J.; Sun, Y.; Chen, Q. GALLERY: A Game-Theoretic Resource Allocation Scheme for Multicell Device-to-Device Communications Underlaying Cellular Networks. *IEEE Internet Things J.* **2015**, *2*, 504–514.

26. Della Penda, D.; Abrardo, A.; Moretti, M.; Johansson, M. Potential games for subcarrier allocation in multi-cell networks with D2D communications. In Proceedings of the 2016 IEEE International Conference on Communications (ICC), Kuala Lumpur, Malaysia, 22–27 May 2016; pp. 1–6.

27. Li, X.Y.; Li, J.; Liu, W.; Zhang, Y.; Shan, H.S. Group-sparse-based joint power and resource block allocation design of hybrid device-to-device and LTE-advanced networks. *IEEE J. Sel. Areas Commun.* **2016**, *34*, 41–57.

28. Jiang, F.; Wang, B.C.; Sun, C.Y.; Liu, Y.; Wang, X. Resource allocation and dynamic power control for D2D communication underlaying uplink multi-cell networks. *Wirel. Netw.* **2016**, pp. 1–15.

29. Foschini, G.J.; Miljanic, Z. A simple distributed autonomous power control algorithm and its convergence. *IEEE Trans. Veh. Technol.* **1993**, *42*, 641–646.

30. Chiang, M.; Hande, P.; Lan, T.; Tan, C.W. Power control in wireless cellular networks. *Found. Trends ® Netw.* **2008**, *2*, 381–533.

31. Papadimitriou, C.H.; Steiglitz, K. *Combinatorial Optimization: Algorithms and Complexity*; Courier Corporation: North Chelmsford, MA, USA, 1982.

32. Ui, T. A Shapley value representation of potential games. *Games Econ. Behav.* **2000**, *31*, 121–135.

33. Kuhn, H.W. The Hungarian method for the assignment problem. *Nav. Res. Logist. Q.* **1955**, *2*, 83–97.

34. 3rd Generation Partnership Project (3GPP). Evolved Universal Terrestrial Radio Access Network (E-UTRAN); X2 General Aspects and Principles. TS 36.420, 2007. Available online: http://www.qtc.jp/3GPP/Specs/36420-800.pdf (accessed on 4 August 2017).
35. Yannakakis, M. Equilibria, fixed points, and complexity classes. *Comput. Sci. Rev.* **2009**, *3*, 71–85.

![future internet logo](future internet) *future internet*

MDPI

Article

Throughput-Aware Cooperative Reinforcement Learning for Adaptive Resource Allocation in Device-to-Device Communication

Muhidul Islam Khan [1,*], Muhammad Mahtab Alam [1], Yannick Le Moullec [1] and Elias Yaacoub [2]

[1] Thomas Johann Seeback Department of Electronics, School of Information Technology, Tallinn University of Technology, Ehitajate tee 5, 19086 Tallinn, Estonia; muhammad.alam@ttu.ee (M.M.A.); yannick.lemoullec@ttu.ee (Y.L.M.)

[2] Faculty of Computer Studies, Arab Open University, Omar Bayhoum Str. - Park Sector, Beirut 2058 4518, Lebanon; eliasy@ieee.org

* Correspondence: mdkhan@ttu.ee; Tel.: +372-5848-8089

Received: 30 September 2017; Accepted: 27 October 2017; Published: 1 November 2017

Abstract: Device-to-device (D2D) communication is an essential feature for the future cellular networks as it increases spectrum efficiency by reusing resources between cellular and D2D users. However, the performance of the overall system can degrade if there is no proper control over interferences produced by the D2D users. Efficient resource allocation among D2D User equipments (UE) in a cellular network is desirable since it helps to provide a suitable interference management system. In this paper, we propose a cooperative reinforcement learning algorithm for adaptive resource allocation, which contributes to improving system throughput. In order to avoid selfish devices, which try to increase the throughput independently, we consider cooperation between devices as promising approach to significantly improve the overall system throughput. We impose cooperation by sharing the value function/learned policies between devices and incorporating a neighboring factor. We incorporate the set of states with the appropriate number of system-defined variables, which increases the observation space and consequently improves the accuracy of the learning algorithm. Finally, we compare our work with existing distributed reinforcement learning and random allocation of resources. Simulation results show that the proposed resource allocation algorithm outperforms both existing methods while varying the number of D2D users and transmission power in terms of overall system throughput, as well as D2D throughput by proper Resource block (RB)-power level combination with fairness measure and improving the Quality of service (QoS) by efficient controlling of the interference level.

Keywords: device-to-device communication; throughput-awareness; cooperative reinforcement learning; system throughput; interference management; adaptive resource allocation

---

1. Introduction

Device-to-device (D2D) communication is a nascent feature for the Long term evolution advanced (LTE-Advanced) systems. D2D communication can operate in centralized, i.e., Base station (BS) controlled mode, and decentralized mode, i.e., without a BS [1]. Unlike the traditional cellular network where Cellular users (CU) communicate through the base station, D2D allows direct communication between users by reusing the available radio resources. Consequently, D2D communication can provide improved system throughput and reduced traffic load to the BS. However, D2D devices generate interferences while reusing the resources [2,3]. Efficient resource allocation play a vital role in reducing the interference level, which positively impacts the overall system throughput. Fine tuning of power allocation on Resource blocks (RB) has consequences on interference, i.e., a higher transmission power can increase D2D throughput; however, it increases the interference level as well. Therefore, choosing the proper level of transmission power for RBs is a key research issue in D2D communication, which calls for adaptive power allocation methods.

Resource allocators, i.e., D2D transmitters in our system model as described in Section 3 need to perform a particular action at each time step based on the application demand. For example, actions can be selecting power level options for a particular RB [4]. Random power allocation is not suitable in a D2D communication due to its dynamic nature in terms of signal quality, interferences and limited battery capacity [5]. Scheduling of these actions associated with different levels of power helps to allocate the resources in such a way that the overall system throughput is increased and an acceptable level of interference is maintained. However, this is hard to maintain, and therefore, we need an algorithm for learning the scheduling of actions adaptively, which helps to improve the overall system throughput with fairness and the minimum level of interferences.

To illustrate the problem, Figure 1 shows a basic single cell scenario with one Cellular user (CU), two D2D pairs and one base station having two resource blocks operating in an underlay mode. D2D devices contend for resource blocks for reusing. Here, RB1 is allocated to the cellular user. D2D pair Tx and D2D pair Rx are assigned RB2. Now, D2D candidate Tx and D2D candidate Rx will contend for the resources either for RB1 or for RB2 to access. If we allocate RB1 to a D2D pair closer to the BS, there will be high interference between the D2D pair and the cellular user. So, RB1 should be allocated to the D2D candidate Tx which is closer to the cell edge (d2 > d3). For reusing the RB1, there will be interferences. Our goal is to propose an adaptive learning algorithm for selecting the proper level of power for the RB to minimize the level of interferences and maximize the throughput of the system.

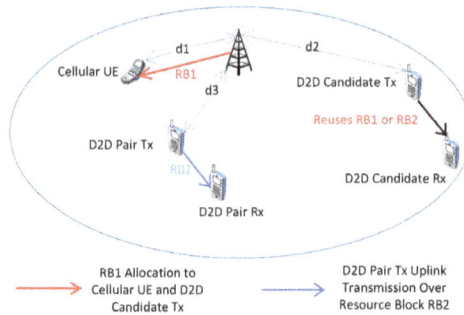

Figure 1. Device-to-device (D2D) communication in a cellular network. RB1 and RB2 resource allocated to the Cellular User equipments (UE) and the D2D pair TX-D2D pair Rx, respectively. D2D candidate Tx-D2D candidate Rx has joined the network, it will contend for the resources, i.e., either reusing RB1 or RB2.

In contrast with existing works, our proposed algorithm helps to learn the proper action selection for resource allocation. We consider reinforcement learning with the cooperation between users by sharing the value function and incorporating a neighboring factor. In addition, we consider a set of states based on system variables which have an impact on the overall QoS of the system. Moreover, we consider both cross-tier interference (interference that the BS receives from D2D transmitter and that the D2D receivers receive from cellular users) and cotier interference (that the D2D receivers receive from D2D transmitters) [6]. To the best of our knowledge, this is the first work that considers all the above aspects for adaptive resource allocation in D2D communications.

The main contributions of this work can be stated as follows:

- We propose an adaptive and cooperative reinforcement learning algorithm to improve achievable system throughput as well as D2D throughput simultaneously. The cooperation is performed by sharing the value function between devices and imposing the neighboring factor in our learning algorithm. A set of actions is considered based on the level of transmission power for a particular Resource block (RB). Further, a set of states is defined considering the appropriate number of system-defined variables. In addition, the reward function is composed of Signal-to-noise-plus-interference ratio (SINR) and the channel gains (between the base station and user, and also between users). Moreover, our proposed reinforcement learning algorithm is an on-policy learning algorithm which considers both exploitation and exploration. This action selection strategy helps to learn the best action to execute, which has a positive impact on selecting the proper level of power allocation to resource blocks. Consequently, this method shows better performance regarding overall system throughput.

- We perform realistic throughput evaluation of the proposed algorithm while varying the transmission power and the number of D2D users. We compare our method with existing distributed reinforcement learning and random allocation of resources in terms of D2D and system throughput considering the system model where Resource block (RB)-power level combination is used for resource allocation. Moreover, we consider fairness among D2D pairs by computing a fairness index which shows that our proposed algorithm achieves balance among D2D users throughput.

The rest of the paper is organized as follows. Section 2 describes the related works. This is followed by the system model in Section 3. The proposed cooperative reinforcement learning based resource allocation algorithm is described in Section 4. Section 5 presents the simulation results. Section 6 concludes the paper with future works.

2. Related Works

Recent advances in Reinforcement learning (RL) create a broad scope of adaptive applications to apply. Resource allocation in D2D communication is such an application. Here, we describe at first some classical approaches [7–16] followed by existing RL-based resource allocation algorithms [17,18].

In [7], an efficient resource allocation technique for multiple D2D pairs is proposed considering the maximization of system throughput. By exploring the relationship between the number of Resource blocks (RB) per D2D pair and the maximum power constraint for each D2D pair, a sub-optimal solution is proposed to achieve higher system throughput. However, the interference among D2D pairs is not considered. Local water filling algorithm (LWFA) is used for each D2D pair which is computationally expensive. Feng et al. [8] introduce a resource allocation technique by maintaining the QoS of cellular users and D2D pairs simultaneously to enhance the system performance. A three-step scheme is proposed where the system performs admission control at first and then allocates the power to each D2D pair and its potential Cellular user (CU). A maximum weight bipartite Matching based scheme (MBS) is proposed to select a suitable CU partner for each D2D pair where the system throughput is maximized. However, this work basically focuses on suitable CU selection for the resource sharing where adaptive power allocation is not considered. In [9], a centralized heuristic approach is proposed where the resources of cellular users and D2D pairs are synchronized considering the interference link gain from D2D transmitter to the BS. They formulate the problem of radio resource allocation to the D2D communication as a Mixed integer nonlinear programming (MINLP). However, MINLP is hard to solve and the adaptive power control mechanism is not considered. Zhao et al. [10] propose a joint mode selection and resource allocation method for the D2D links to enhance the system sum-rate. They formulate the problem to maximize the throughput with SINR and power constraints for both D2D links and cellular users. They propose a Coalition formation game (CFG) with transferable utility to solve the problem. However, they do not consider the adaptive power allocation problem. In [11], Min et al. propose a Restricted interference region (RIR) where cellular users and D2D users can not coexist. By adjusting the size of the restricted interference region, they propose the interference control mechanism in a way that the D2D throughput is increased over time. In [12], the authors consider the target rate of cellular users for maximizing the system throughput. Their proposed method shows better results in terms of system interference. However, their work also focuses on the region control for the interference. They do not consider the adaptive resource allocation for maximizing the system throughput. A common limitation to the works as mentioned above is that they are fully centralized, which requires full knowledge of the link state information that produces redundant information over the network.

In addition to above-mentioned works, Hajiaghajani et al. [13] propose a heuristic resource allocation method. They design an adaptive interference restricted region for the multiple D2D pairs. In their proposed region, multiple D2D pairs share the resources where the system throughput is increased. However, their proposed method is not adaptive regarding power allocation to the users. In [14], the authors propose a two-phase optimization algorithm for the adaptive resource allocation which provides better results for system throughput. They propose Lagrangian dual decomposition (LDD) which is computationally complex.

Wang et al. [15] propose a Joint scheduling (JS) and resource allocation for the D2D underlay communication where the average D2D throughput can be improved. Here, the channel assigned to the cellular users is reused by only one D2D pair and the cotier interference is not considered. In [16], Yin et al. propose a distributed resource allocation method where minimum rates of cellular users and D2D pairs are maintained. A Game theoretic algorithm (GTA) is proposed for minimizing the interferences among D2D pairs. However, this approach provides low spectral efficiency.

With regards to machine learning for resource allocation in D2D communication, there are only few works, e.g., [17,18]. Luo et al. [17] and Nie et al. [18] exploit machine learning algorithms for D2D resource allocation. Luo et al. [17] propose Distributed reinforcement learning (DIRL), Q-learning algorithm for resource allocation which improves the overall system performance in comparison to the random allocator. However, the model of Reinforcement learning (RL) is not well structured. For example, the set of states and a set of actions are not adequately designed. Their reward function is composed of only Signal to interference plus noise power ratio (SINR) metric. The channel gain between the base station and the user, and also the channel gain between users are not considered. This is a drawback since channel gains are important to consider as these help the D2D communication with better SINR level and transmission power, which is reflected in increased system throughput [19].

Recently, Nie et al. [18] propose Distributed reinforcement learning (DIRL), Q-learning to solve the power control problem in underlay mode. In addition, they explore the optimal power allocation which helps to maintain the overall system capacity. However, this preliminary study has limitations, for example, in their reward function, the channel gains are not considered. In addition, in their system model only the transmit power level is considered for maximizing the system throughput. To consider RB/subcarrier allocation in the optimization function is a very important issue for mitigating interference [20]. Moreover, the cooperation between devices for resource allocation is not investigated in these existing works. A summary of the features and limitations of classical and RL-based allocation methods is given in Table 1.

We propose adaptive resource allocation using Cooperative reinforcement learning (CRL) considering the neighboring factor, improved state space, and a reward function. Our proposed resource allocation method helps to provide mitigated interference level, D2D throughput and consequently an overall improved system throughput.

Table 1 shows the comparison of all the above mentioned works with our proposed cooperative reinforcement learning. Firstly, we categorize the related methods in two types: classical D2D resource allocation methods and Reinforcement learning (RL) based D2D resource allocation methods. We compare these works based on D2D throughput, system throughput, transmission alignment, online task scheduling for resource allocation, and cooperation. We can observe that almost all the methods consider the D2D and system throughput. None of the existing methods for resource allocation consider the transmission alignment, online action scheduling, and cooperation for the adaptive resource allocation.

Table 1. Comparison of existing methods with the proposed cooperative reinforcement learning.

Methods	References	D2D Throughput	System Throughput	Transmission Alignment	Action Scheduling	Cooperation
Classical approaches	LWFA [7]	Yes	Yes	No	N/A	No
	MBS [8]	Yes	Yes	No	N/A	No
	MINLP [9]	Yes	No	No	N/A	No
	CFG [10]	Yes	No	No	N/A	No
	RIR [11]	Yes	No	No	N/A	No
	RIR [12]	Yes	No	No	N/A	No
	RIR [13]	Yes	No	No	N/A	No
	LDD [14]	Yes	No	No	N/A	No
	JS [15]	Yes	Yes	No	N/A	No
	GTA [16]	Yes	Yes	No	N/A	No
RL based method	DIRL [17]	Yes	No	No	No	No
	DIRL [18]	No	Yes	No	No	No
Proposed method	Cooperative RL	Yes	Yes	Yes	Yes	Yes

3. System Model

We consider a network that consists of one Base station (BS) and a set of $\check{C}$ Cellular users (CU), i.e., $\check{C} = \{1,2,3,\ldots,C\}$. There are also $\check{D}$ D2D pairs, $\check{D} = \{1,2,3,\ldots,D\}$ coexist with the cellular users within the coverage of BS. In a particular D2D pair, d_T and d_R are the D2D transmitter and D2D receiver respectively. The set of User equipments (UE) in the network is given by UE = $\{\check{C} \cup \check{D}\}$. Each D2D transmitter d_T selects an available Resource block (RB) r from the set $RB = \{1,2,3,\ldots,R\}$. In addition, underlay D2D transmitters select the transmit power from a finite set of power levels, i.e., $p_r = (p_r^1, p_r^2, \ldots, p_r^R)$. Each D2D transmitter should select resources, i.e., RB-power level combination refers to transmission alignment [21].

For each RB $r \in R$, there is a predefined threshold $I_{th}^{(r)}$ for maximum aggregated interference. We consider that the value of $I_{th}^{(r)}$ is known to the transmitters using the feedback control channels. An underlay transmitter uses a particular transmission alignment in a way that the cross-tier interference should be within the threshold limit. According to our proposed system model, only one CU can be served by one RB where D2D users can reuse the same RB to improve the spectrum efficiency.

For each transmitter d_T, the transmit power over the RBs is determined by the vector $p_r = [p_r^1, p_r^2, \ldots, p_r^R]^T$ where $p_r \geq 0$ denotes the transmit power level of transmitter over resource block r. If RB is not allocated to the transmitter then the power level $p_r = 0$. As we assume that each transmitter selects only one RB where only one entity in the power level $p_r \neq 0$.

Signal-to-interference-plus-noise-ratio (SINR) can be treated as an important factor to measure the link quality. The received SINR for any D2D receiver over rth RB as follows:

$$\gamma_r^{D_u} = \frac{p_r^{D_u}.G_{D_{u,r}}^{uu}}{\sigma^2 + p_r^c.G_r^{cu} + \sum_{v \in D_r}^{v \neq u} p_r^{d_v}.G_{D_{v,r}}^{uv}} \tag{1}$$

where $p_r^{D_u}$ and p_r^c denote the uth D2D user and cellular user uplink transmission power on rth RB, respectively. $p_r^{D_u} \leq P_{max}, \forall u \in D$ where P_{max} is the upper bound of each D2D user's transmit power. σ^2 is the noise variance [9].

$G_{D_{u,r}}^{uu}$, $G_{D_{v,r}}^{uv}$ and G_r^{cu} are the channel gains in the uth D2D link, the channel gain from D2D transmitter u to receiver v, and the channel gain from cellular transmitter c to receiver u, respectively. D_r is a D2D pairs set sharing the rth RB.

The SINR of a cellular user $c \in \check{C}$ on the rth RB is

$$\gamma_r^c = \frac{p_r^c.G_{c,r}}{\sigma^2 + \sum_{v \in D_r} p_r^{d_v} G_{v,r}} \tag{2}$$

where $G_{c,r}$ and $G_{v,r}$ indicate the channel gains on the rth RB from BS to cellular user c and vth D2D transmitter, respectively.

The total path-loss which includes the antenna gain between BS and the user u is:

$$PL_{dB,B,u(.)} = L_{dB}(d) + \log_{10}(X_u) - A_{dB}(\theta) \tag{3}$$

where $L_{dB}(d)$ is the pathloss between a BS and the user at a distance d meter. X_u is the log-normal shadow path-loss of user u. $A_{dB}(\theta)$ is the radiation pattern [22].

$L_{dB}(d)$ can be expressed as follows:

$$L_{dB}(d) = 40(1 - 4 \times 10^{-3}h_b)\log_{10}(d/1000)$$
$$-18\log_{10}(h_b) + 21\log_{10}(f_c) + 80 \tag{4}$$

where f_c is the carrier frequency in GHz and h_b is the base station antenna height [22].

The linear gain between the BS and a user is $G_{Bu} = 10^{\frac{-PL_{dB,B,u}}{10}}$.

For D2D communication, the gain between two users u and v is $G_{uv} = k_{uv}d_{uv}^{-\alpha}$ [23]. Here, d_{uv} is the distance between transmitter u and receiver v. α is a constant pathloss exponent and k_{uv} is a normalization constant.

The objective of resource allocation problem (i.e., to allocate RB and transmit power) is to assign the resources in a way that maximizes system throughput. System throughput is the sum of D2D users and CU throughput, which is calculated by Equation (6).

The resource allocation can be indicated by a binary decision variable, $b_v^{(r,p_r)}$ where

$$b_v^{(r,p_r)} = \begin{cases} 1, & \text{if the transmitter } v \text{ is transmitting over RB } r \text{ with power level } p_r \\ 0, & \text{otherwise} \end{cases}$$

The aggregated interference experienced by RB r can be expressed as follows

$$I^{(r)} = \sum_{v=1}^{\check{D}} \sum_{p_r=1}^{P_{max}} b_v^{(r,p_r)} G_{v,r} p_r \tag{5}$$

Let $B = [b_1^{(1,1)}, \ldots, b_1^{(r,p_r)}, \ldots, b_1^{(R,P_{max})}]^T$ denote the resource e.g., RB and transmission power allocation. So, the allocation problem can be expressed as follows:

$$\max_B \sum_{r=1}^{R} \sum_{p_r=1}^{P_{max}} b_v^{(r,p_r)} W_{RB}\{\log_2(1+\gamma_r^c) + \sum_{u \in D_r} \log_2(1+\gamma_r^{D_u})\}$$
$$\text{subject to } I^{(r)} < I_{th}^{(r)}, \forall_r$$
$$b_v^{(r,l)} \in \{0,1\}, \forall_{v,r,l} \tag{6}$$
$$\sum_{r=1}^{R} \sum_{p_r=1}^{P_{max}} b_v^{(r,l)} = 1, \forall_v$$
$$0 \le p_r \le P_{max}, \forall_{u,r}$$

where $p_r = (p_r^1, p_r^2, \ldots, p_r^R)$ and W_{RB} is the bandwidth corresponding to a RB. The objective function is to maximize the throughput of the system constrained by that the aggregated interference should be limited by a predefined threshold. The number of RB selected by the transmitter should be one where each can select one power level at each RB. Our goal is to investigate the optimal resource allocation in such a way that the system throughput is maximized by applying cooperative reinforcement learning.

Future Internet **2017**, *9*, 72

4. Cooperative Reinforcement Learning Algorithm for Resource Allocation

In this section, we describe the basics of Reinforcement learning (RL), followed by our proposed cooperative reinforcement learning algorithm. After that, we describe the set of states, the set of actions and reward function for our proposed algorithm. Finally, Algorithm 1 shows the overall proposed resource allocation method and Algorithm 2 shows the execution steps of our proposed cooperative reinforcement learning.

We apply a Reinforcement learning (RL) algorithm named state action reward state action, SARSA(λ), for adaptive resource in D2D communication for efficient resource allocation. This variant of standard SARSA(λ) [24] algorithm has some important features like cooperation by using a neighboring factor, a heuristic policy for exploration and exploitation, and a varying learning rate considering the visited state-action pair. Currently, we are applying the learning algorithm for the resource allocation of D2D users considering that the allocation of cellular users is performed prior to the allocation of D2D users. We consider the cooperative fashion of this learning algorithm which helps to improve the throughput as explained in Section 1 by sharing the value function and incorporating weight factors for the neighbors of each agent.

In reinforcement learning, there is no need for prior knowledge about the environment. Agents learn how to behave with the environment based on the previous experience achieved, which is traced by a parameter, i.e., Q-value and controlled by a reward function. There should be some actions/tasks to perform at every time step. After performing every action, the agents shifts from one state to another and it gets a reward that reflects the impact of that action, which helps to decide about the next action to perform. The basic reinforcement learning is a form of Markov decision process (MDP).

Figure 2 depicts the overall model of a reinforcement learning algorithm.

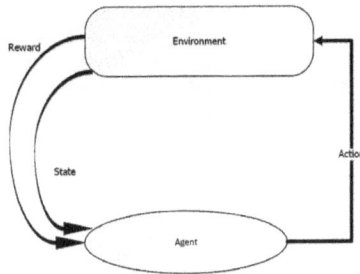

Figure 2. Basic components of a reinforcement learning. The agent performs an action to the environment which gives a reward and helps to shift from one state to another.

Each agent in RL has the following components [25]:

- Policy: The policy acts as a decision making function for the agents. All other functions/components help to improve the policy for better decision making.
- Reward function: The reward function defines the ultimate goal of an agent. This helps to assign a value/number to the performed action, which indicates the intrinsic desirability of the states. The main objective of the agent is to maximize the reward function in the long run.
- Value function: The value function determines the suitability of action selection in the long run. The value of a state is accumulated reward over long run when starting from the current state.
- Model: The model of the environment mimics the behavior of the environment which consists of a set of states and a set of actions.

In our model of the environment, we consider the components of the reinforcement learning algorithm as follows:

Agent: All the resource allocators: D2D Transmitters.

State: The state of D2D user u on RB r at time t is defined as:

$$S_t^{u,r} = \gamma_r^c \cup G_{Bu} \cup G_{uv}$$

We consider three variables γ_r^c, G_{Bu} and G_{uv} for defining the states for maintaining the overall quality of the network. γ_r^c is the SINR of a cellular user on the rth RB. G_{Bu} is the channel gain between the BS and an user u. G_{uv} is the channel gain between two users u and v. The variables γ_r^c, G_{Bu} and G_{uv} are important to consider for the resource allocation. The SINR γ_r^c is the indicator of the quality of service of the network. In addition, if the channel gains (G_{Bu} and G_{uv}) quality is good then it is possible to achieve higher throughput without excessively increasing the transmit power, i.e., without causing too much interference to others. On the other hand, if the channel gain is too low, higher transmit power is required, which leads to increased interference.

Now, the state values of these variables can be either 0 or 1 based on following conditions. If the value of the variables are greater than or equal to a threshold value, then this denotes that their state value is '0'. On the contrary, if the values are less than the threshold value, then their state value is '1'. So, $\gamma_r^c \geq \tau_0$ means state value '1' and $\gamma_r^c < \tau_0$ means state value '0'. Similarly, $G_{Bu} \geq \tau_1$ means state value '1' and $G_{Bu} < \tau_1$ means state value '0'. Consequently, $G_{uv} \geq \tau_2$ means state value '1' and $G_{uv} < \tau_2$ means the state value '0'. In this way, based on the combination of the value of these variables, the total number of possible states is eight where τ_0, τ_1 and τ_2 are the minimum SINR and channel gain guaranteeing the QoS performance of the system.

Action/Task: The action of each agent consists of a set of transmitting power levels. It is denoted by

$$A = (a_r^1, a_r^2, \dots, a_r^{pl})$$

where r represents the rth Resource Block (RB), and pl means that every agent has pl power levels. In this work, we consider the power levels to assign within the range of 1 to P_{max} in the interval of 1 dBm.

Reward Function: The reward function for the reinforcement learning is designed focusing on the throughput of each agent/user which is formulated as follows:

$$\Re = \log_2(1 + SINR(u)) \tag{7}$$

when $\gamma_r^c \geq \tau_0$, $G_{Bu} \geq \tau_1$ and $G_{uv} \geq \tau_2$. Otherwise, $\Re = -1$. SINR (u) denotes the signal to interference plus noise power ratio of user u (Step 7–10 in Algorithm 1).

SARSA(λ) is an on-policy reinforcement learning algorithm that estimates the value of the policy being followed where λ is a parameter such as learning rate [26]. In SARSA learning algorithm, every agent needs to maintain a Q matrix which is initially assigned 0 and the agents may be in any state. Based on performing one particular action, it shifts from one state to another. The basic form of the learning algorithm is $(s_t, a_t, \Re, s_{t+1}, a_{t+1})$, which means that the agent was in state s_t, did action a_t, received reward $\Re$, and ended up in state s_{t+1}, from which it decided to perform action a_{t+1}. This provides a new iteration to update $Q_t(s_t, a_t)$.

SARSA(λ) helps to find out the appropriate sets of actions for some states. The considered state-action pair's value function $Q_t(s_t, a_t)$ as follows:

$$Q_t(s_t, a_t) = \Re + \gamma Q_{t+1}(s_{t+1}, a_{t+1}) \tag{8}$$

In Equation (8), γ is a *discount-factor* which varies from 0 to 1. The higher the value, the more the agent relies on future rewards than on the immediate reward. The objective of applying reinforcement learning is to find the optimal policy $Q_t^\pi(s_t, a_t)$ which maximizes the value function $\pi = \max_\pi Q_t^\pi(s_t, a_t)$.

We consider the cooperative fashion of this algorithm where each agent shares the value function with each other.

At each time step, Q_{t+1} for the iteration $t+1$, Q_{t+1} is updated with the temporal difference error δ_t and the immediate received reward. The Q value has the following update rules:

$$Q_{t+1}(s_{t+1}, a_{t+1}) \leftarrow Q_t(s_t, a_t) + \alpha \delta_t e_t(s_t, a_t) \tag{9}$$

for all s, a.

In Equation (9), $\alpha \in [0, 1]$ is the learning rate which decreases with time. δ_t is the temporal difference error which is calculated by following rule (Step 7 in Algorithm 2):

$$\delta_t = \Re + \gamma_1 f Q_{t+1}(s_{t+1}, a_{t+1}) - Q_t(s_t, a_t) \tag{10}$$

In Equation (10), γ_1 is a discount-factor which varies from 0 to 1. The higher the value, the more the agent relies on future rewards than on the immediate reward. $\Re_{t+1}$ represents the reward received for performing an action. f is the neighboring weight factor of agent i where this factor consists of the effect of neighbor's Q-value, which helps to update the Q-value of agent i that is calculated as follows [27]:

$$f = \frac{1}{ngh(n_i)} \quad \text{if } ngh(n_i) \neq 0 \tag{11}$$

$$= 1 \quad \text{otherwise.} \tag{12}$$

where $ngh(n_i)$ is the number of neighbors of agent i within the D2D radius. BS provides the information of number of neighbors for each agent [28].

There is a trade-off between exploration and exploitation in reinforcement learning. Exploration chooses an action randomly in the system to find out the utility of that chosen action. Exploitation deals with the actions which have been chosen based on previously learned utility of the actions.

We use a heuristic for exploration probability at any given time such as:

$$\epsilon = \min(\epsilon_{max}, \epsilon_{min} + k * (S_{max} - S)/S_{max}) \tag{13}$$

where ϵ_{max} and ϵ_{min} denote upper and lower boundaries for the exploration factor, respectively. S_{max} represents the maximum number of states which is eight in our work and S represents the current number of states already known [29]. At each time step, the system calculates ϵ and generates a random number in the interval $[0, 1]$. If the selected random number is less than or equal to ϵ, the system chooses a uniformly random task (exploration), otherwise it chooses the best task using Q values (exploitation). k is a constant which controls the effect of unexplored states (Step 4 in Algorithm 2).

SARSA(λ) helps to improve the learning technique by eligibility trace. In Equation (9), $e_t(s, a)$ is the eligibility trace. The eligibility trace is updated by the following rule:

$$e_t(s_t, a_t) = \gamma_2 \lambda e_{t-1}(s_t, a_t) + 1 \quad \text{if } s_t \in s \text{ and } a_t \in a$$
$$e_t(s_t, a_t) = \gamma_2 \lambda e_{t-1}(s_t, a_t) \quad \text{otherwise.}$$

Here, λ is learning parameter for guaranteed convergence, whereas γ_2 is the discount factor. In addition, the eligibility trace helps to provide higher impact on revisited states. For example, for a state-action pair (s_t, a_t), if $s_t \in s$ and $a_t \in a$, the state-action pair is reinforced. Otherwise, the eligibility trace is removed (Step 8 in Algorithm 2).

The learning rate α is decreased in such a way that it reflects the degree to which a state-action pair has been chosen in the recent past. It is calculated as:

$$\alpha = \frac{\rho}{visited(s,a)} \tag{14}$$

where ρ is a positive constant and $visited(s,a)$ represents the visited state-action pairs so far [30] (Step 6 in Algorithm 2).

Algorithm 1: Proposed resource allocation method

    Input : P_{max} = 23 dBm, Number of resource blocks = 30, Number of cellular users = 30, Number of D2D user pairs = 12, D2D radius = 20 m, Pathloss parameter = 3.5, Cell radius = 500 m, τ_0 = 0.004, τ_1 = 0.2512, τ_2 = 0.2512 [9]

    Output: RB-Power level, System Throughput

1 **loop**

2 Pathloss calculation by $PL_{dB,B,u(.)} = L_{dB}(d) + log_{10}(X_u) - A_{dB}(\theta)$

3 Gain between the BS and a user, $G_{Bu} = 10^{\frac{-PL_{dB,B,u}}{10}}$

4 Gain between two users, $G_{uv} = k_{uv}d_{uv}^{-\alpha}$

5 SINR of the D2D users on the rth RB, $\gamma_r^{Du} = \frac{p_r^{Du}.G_{Du,r}^{uu}}{\sigma^2 + p_r^c.G_r^{cu} + \sum_{v \in D_r}^{v \neq u} p_r^{dv}.G_{Dv,r}^{uv}}$

6 SINR of the cellular users on the RB, $\gamma_r^c = \frac{p_r^c.G_{c,r}}{\sigma^2 + \sum_{v \in D_r} p_r^{dv} G_{v,r}}$

7 **if** ($\gamma_r^c \geq \tau_0$, $G_{Bu} \geq \tau_1$ and $G_{uv} \geq \tau_2$) **then**

8 | $\Re = log_2(1 + SINR(u))$;

9 **else**

10 | $\Re = -1$;

11 **end**

12 Apply Algorithm 2 for the power allocation

13 **end loop**

Algorithm 2: Cooperative SARSA(λ) reinforcement learning algorithm over number of iterations.

---

1 Initialize $Q(s,a) = 0$, $e(s,a) = 0$, $\epsilon_{max} = 0.3$, $\epsilon_{min} = 0.1$, $k = 0.25$, $\rho = 1$, $\gamma = 0.9$, $\gamma_1 = 0.5$, $\lambda = 0.5$ [17,29]

2 **loop**

3 Determine the current s based on γ_r^c, G_{Bu} and G_{uv}

4 Select a particular action a based on the policy, $\epsilon = min(\epsilon_{max}, \epsilon_{min} + k * (S_{max} - S) / S_{max})$

5 Execute the selected action

6 Update learning rate by $\alpha = \frac{\rho}{visited(s,a)}$

7 Determine the temporal difference error by $\delta_t = \Re + \gamma_1 f Q_{t+1}(s_{t+1}, a_{t+1}) - Q_t(s_t, a_t)$

8 Update eligibility traces

9 Update the Q-value, $Q_{t+1}(s_{t+1}, a_{t+1}) \leftarrow Q_t(s_t, a_t) + \alpha \delta_t e_t(s_t, a_t)$

10 Update the value function and share with neighbors

11 Shift to the next based on the executed action

12 **end loop**

Algorithm 1 depicts the overall proposed resource allocation method. After setting the initial input parameters, the system oriented parameters, i.e., pathloss, channel gains, SINR of the D2D users and cellular users on the rth RB are calculated (Step 2–6 in Algorithm 1). Then the reward function is calculated (Step 8) and is assigned when the state values satisfy the constraint in step 7.

After that Algorithm 2 is applied for the adaptive resource allocation. Algorithm 2 shows our proposed reinforcement learning algorithm execution steps for resource allocation over number of iterations.

5. Performance Evaluation

We implement our proposed cooperative reinforcement learning algorithm and compare it with the random allocation and existing distributed reinforcement learning algorithm.

The parameters for the simulation are shown in Table 2.

Table 2. Simulation Parameters.

Parameter	Value
P_{max}	23 dBm
Number of resource blocks	30
Number of cellular users	30
Number of D2D user pairs	12
D2D radius	20 m
Pathloss parameter	3.5
Cell radius	500 m
τ_0	0.004
τ_1	0.2512
τ_2	0.2512
$I_{th}^{(r)}$	0.001
W_{RB}	180 kHz
Initial $Q(s,a)$	0
Initial $e(s,a)$	0
ϵ_{max}	0.3
ϵ_{min}	0.1
k	0.25
ρ	1
γ	0.9
γ_1	0.5
λ	0.5

We consider a single cell with a radius of 500 m where some cellular users and D2D pairs are uniformly distributed within the coverage of the BS. There are 30 cellular users and 12 D2D users. We consider a constraint of resources with only 30 resource blocks.

We consider $\tau_0 = 0.004$, $\tau_1 = 0.2512$ and $\tau_2 = 0.2512$ as constraints to define the states for maintaining the quality of service [9]. In our reinforcement learning algorithm, we consider $\epsilon_{max} = 0.3$, $\epsilon_{min} = 0.1$ and $k = 0.25$ [29]. We set $\rho = 1$ for learning rate calculation in Equation (14). The discount factor, $\gamma_1 = 0.9$ is considered based on the work [17] for fair comparison with our work.

We compare our method with the distributed reinforcement learning proposed in [17] and a base-line random allocation of resources.

5.1. Throughput Analysis over Number of Iterations

Figure 3a shows that the proposed cooperative reinforcement learning outperforms both the random allocation and the distributed reinforcement learning regarding average system throughput calculated by Equation (6) considering 12 D2D user pairs and other parameter values as in Table 2. We can observe that after the 30th iteration (Figure 3b), our proposed learning algorithm outperforms other methods at almost every iteration when the algorithm reaches the convergence of learning. In addition, there are variations in throughput results for each method and also there are some points where distributed reinforcement learning outperforms proposed cooperative reinforcement learning due to the fact that we consider the heuristic action selection policy based on exploration and exploitation in Equation (13), which avoids to stuck the learning algorithm in a local optimum. Random allocation shows poor results since it does not act appropriately with the changes of the environment. Whereas, distributed reinforcement learning shows moderate results comparing with the both methods. We consider 100 iterations here for the comparison, but the trend of outperformance of our proposed algorithm remains the same with additional iterations.

Figure 3b shows the Q-values calculated by Equation (9) of distributed reinforcement learning and cooperative reinforcement learning over number of iterations for learning. We can observe that the cooperative reinforcement learning converges faster due to the sharing learned policies between devices. Further, our proposed learning algorithm provides better Q-value at each time step which imposes an impact on the overall improved system throughput. Moreover, higher Q-values denote that action scheduling strategies are performed much better in our proposed algorithm.

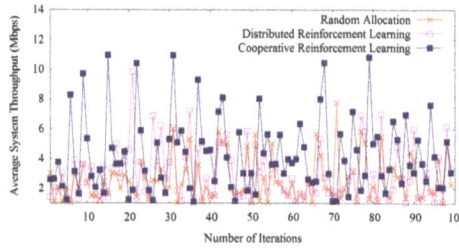

(a) Average system throughput over number of iterations

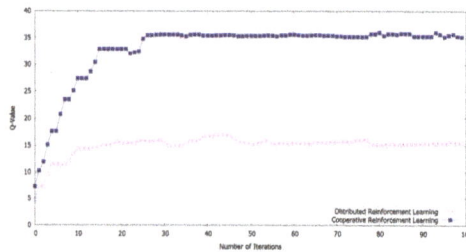

(b) Convergence of learning algorithms

Figure 3. (a) Average system throughput over number of iterations (b) Convergence of learning algorithms.

5.2. Throughput Analysis by Varying the Transmit Power Level

Figure 4 shows the average D2D throughput over transmit power applying our proposed method, distributed reinforcement learning and random allocation of resources. All the methods follow the same trend that with the increase of transmit power, D2D throughput increases. Our proposed reinforcement learning outperforms others at every level of transmit power due to the appropriate learning of transmission power assignment to the resource blocks. Our proposed method increases D2D throughput by 6.2% as compared with the distributed reinforcement learning. On the other hand, random allocation shows the lowest performance as usual due to the allocation of resources without adaptiveness.

Figure 5 shows the trade-off between D2D throughput and cellular user throughput when varying the transmit power to these values {0.0569, 0.0741, 0.0800} after applying the proposed cooperative reinforcement learning, random allocation, and distributed reinforcement learning where the results are grouped into three sets for each method considering 12 D2D users. For all methods, we can observe the same trends, for example, with the increase of transmit power; D2D throughput increases but cellular user throughput decreases which show the typical phenomena of D2D communication. When the transmit power of the D2D device increases, the D2D throughput also increases. But this provides an impact of interference level to the cellular users which provides lower cellular user throughput. Our proposed method outperforms all methods in terms of D2D and cellular user throughput. For example, when the transmit power is equal to 0.0569, our proposed learning algorithm provides a D2D throughput equal to 3.80, and the cellular user throughput is equal to 2.912. On the other hand, for distributed reinforcement learning, the D2D and the cellular user throughput are 3.53 and 2.734, respectively which is lower than our proposed method. Moreover, random allocation of resources provides the least amount of D2D and cellular user throughput with values equal to 2.62 and 2.27, respectively.

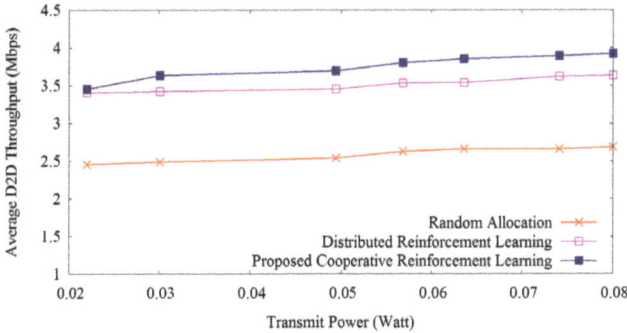

Figure 4. D2D throughput versus transmit power.

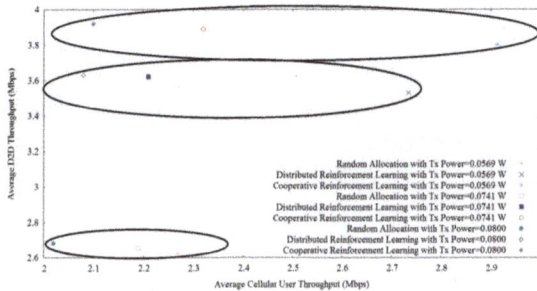

Figure 5. Joint D2D throughput and cellular user throughput optimization.

5.3. Throughput Analysis over a Number of D2D Users

Figure 6 shows the average D2D throughput, average cellular user throughput and the average system throughput over the number of D2D users. We can observe that D2D throughput increases with the increase of D2D users, but on the other hand, cellular user throughput decreases. System throughput is the summation of D2D and the cellular user throughput, which also increases with the increment of the D2D users. For example, our proposed method provides a cellular user throughput equal to 2.912 for 10 D2D users. The proposed method yields 0.2880 as D2D user throughput, which gives a system throughput equal to 3.2. When we increase the number of D2D users to 20, the figure shows a cellular user throughput, D2D throughput, and overall system throughput of 2.8259, 0.5477 and 3.3736, respectively.

All methods show these same trends over the number of D2D users. From this experiment, we can also investigate the issue about the appropriate number of D2D users which provides the better trade-off between D2D and cellular user throughput in a single cell scenario. Here, we can observe that moderate number of D2D users, for example, 50 D2D users provide suitable amount of D2D and cellular user throughput. Our proposed method outperforms the other methods regarding D2D throughput, cellular user throughput and overall system throughput at every number of D2D users.

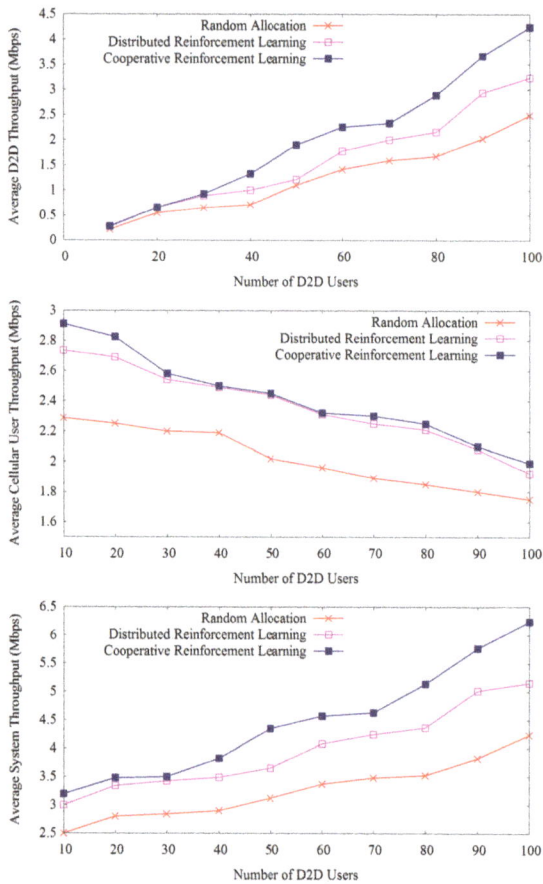

Figure 6. Throughput analysis over a number of D2D users.

5.4. Fairness Analysis

We compute the fairness of our proposed algorithm, the distributed reinforcement learning algorithm, and the random allocation using Jain's fairness index [31]. Jain's fairness index can be derived as

$$f(D_1, D2, \ldots, D_n) = \frac{(\sum_{i=1}^{N} D_i)^2}{N \sum_{i=1}^{N} D_i^2}$$

where D is the throughput of each device and N is the number of users. Jain's fairness is used to determine the fairness of the algorithm which helps to make a stable environment of D2D throughput for each D2D pairs.

Figure 7 shows the fairness measure of our proposed learning algorithm, distributed reinforcement learning and random allocation of resources. The higher value of Jain's index shows a better balance of resources, i.e., fairness. We observe that our proposed algorithm outperforms the two others in terms of fairness. With the increase of D2D pairs, we can see that there is a slight decrease in the fairness level. As our learning algorithm works online based on the proper level of transmit power allocation, we get suitable results in terms of fairness. The distributed reinforcement learning shows comparably less performance with the proposed method. Random allocation of resource provides the worst fairness because it does not use adaptive action scheduling strategy for the resource allocation.

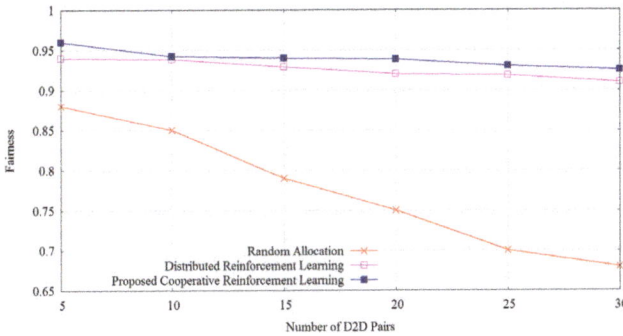

Figure 7. Fairness index of D2D throughput versus number of D2D pairs.

6. Conclusions

Adaptive resource allocation is a critical issue for the current context in D2D communication. Reinforcement learning can be considered as a suitable method for the adaptive resource allocation by scheduling the actions performed by the resource allocators. In this work, we apply a cooperative reinforcement learning for the resource allocation to improve the system throughput and D2D throughput. A key aspect of our work is that we consider cooperation between agents by sharing the value function and imposing a neighboring factor. In addition, the set of states for our proposed reinforcement learning is composed of important system defined variables which helps to increase the observation space that has to be explored. We measure the fairness of our proposed algorithm considering a fairness index. Our method is compared with the distributed reinforcement learning and random allocation of the resource. The results show better performance regarding system throughput, D2D throughput, and the fairness measure.

All results (showed in Section 5) help us to reach the following decisions:

- Our proposed cooperative reinforcement learning method provides better performance regarding system throughput compared to the distributed reinforcement learning, and random allocation of resources. There are some time steps where distributed reinforcement learning outperforms our proposed method due to our heuristic action selection strategy for exploration and exploitation.
- Our proposed method outperforms the distributed reinforcement learning and random allocation of resources in terms of D2D throughput while varying the transmit power. It is possible to observe that in our proposed method, D2D throughput increases about 6.2% compared to the distributed reinforcement learning.
- The trade-off is observed for D2D and cellular user throughput by varying the transmit power at different values, we can observe that higher transmit power provides higher D2D throughput. Our proposed reinforcement learning provides better results regarding both D2D and cellular user throughput compared to the distributed reinforcement learning and random allocation of resources. By increasing the number of D2D users, we can observe higher D2D and the system throughput.
- Higher index value provides higher fairness measure in Jain's fairness index. We can observe that our proposed reinforcement learning outperforms the distributed reinforcement learning and random allocation of resources regarding fairness measure.

Currently, we are considering a single cell for our work. As future work, we will consider multiple cells with more dynamics in the environment. Investigating the latency and energy efficiency issues can further be considered. In addition, currently we are not considering the learning algorithms for BS. To include BS as an agent to apply learning algorithm might enhance the QoS for allocating resources to cellular users. Designing the system in a more distributed way considering multiple cells might utilize the full benefits of applying reinforcement learning.

Acknowledgments: This research was supported by the Estonian Research Council through the Institutional Research Project IUT19-11, and by the Horizon 2020 ERA-chair Grant "Cognitive Electronics COEL"-H2020-WIDESPREAD-2014-2 (Agreement number: 668995; project TTU code VFP15051).

Author Contributions: The main concept to solve the problem was conceived by Muhidul Islam Khan. Furthermore, the initial system model was designed and implemented by Muhidul Islam Khan. Muhammad Mahtab Alam worked on to further improve the problem formulation, overall concept of the work and to improve the writing structure of the paper. Elias Yaacoub provided useful suggestions based on to improve the system model and basic components of the method. Yannick Le Moullec helped to enhance the writing style, and review the paper to make it more structured.

Conflicts of Interest: The authors declare no conflict of interest.

References

1. Doppler, K.; Rinne, M.; Wijting, C.; Ribeiro, C.B.; Hugl, K. Device-to-device communication as an underlay to LTE-advanced networks. *IEEE Commun. Mag.* **2009**, *47*, doi:10.1109/MCOM.2009.5350367.
2. Fodor, G.; Dahlman, E.; Mildh, G.; Parkvall, S.; Reider, N.; Miklós, G.; Turányi, Z. Design aspects of network assisted device-to-device communications. *IEEE Commun. Mag.* **2012**, *50*, doi:10.1109/MCOM.2012.6163598.
3. Xiao, X.; Tao, X.; Lu, J. A QoS-aware power optimization scheme in OFDMA systems with integrated device-to-device (D2D) communications. In Proceedings of the 2011 IEEE Vehicular Technology Conference (VTC Fall), San Francisco, CA, USA, 5–8 September 2011; IEEE: Piscataway, NJ, USA, 2011; pp. 1–5.
4. Khan, M.I.; Rinner, B. Resource coordination in wireless sensor networks by cooperative reinforcement learning. In Proceedings of the 2012 IEEE International Conference on Pervasive Computing and Communications Workshops (PERCOM Workshops), Lugano, Switzerland, 19–23 March 2012; IEEE: Piscataway, NJ, USA, 2012; pp. 895–900.
5. Della Penda, D.; Fu, L.; Johansson, M. Energy efficient D2D communications in dynamic TDD systems. *IEEE Trans. Commun.* **2017**, *65*, 1260–1273.
6. Boabang, F.; Nguyen, H.H.; Pham, Q.V.; Hwang, W.J. Network-Assisted Distributed Fairness-Aware Interference Coordination for Device-to-Device Communication Underlaid Cellular Networks. *Mob. Inf. Syst.* **2017**, *2017*, 1821084.
7. Kai, Y.; Zhu, H. In Proceedings of the Resource allocation for multiple-pair D2D communications in cellular networks. In Proceedings of the 2015 IEEE International Conference on Communications (ICC), London, UK, 8–12 June 2015, IEEE: Piscataway, NJ, USA, 2015; pp. 2955–2960.
8. Feng, D.; Lu, L.; Yuan-Wu, Y.; Li, G.Y.; Feng, G.; Li, S. Device-to-device communications underlaying cellular networks. *IEEE Trans. Commun.* **2013**, *61*, 3541–3551.
9. Zulhasnine, M.; Huang, C.; Srinivasan, A. Efficient resource allocation for device-to-device communication underlaying LTE network. In Proceedings of the 2010 IEEE 6th International Conference on Wireless and Mobile Computing, Networking and Communications (WiMob), Niagara Falls, NU, Canada, 11–13 October 2010; IEEE: Piscataway, NJ, USA, 2010; pp. 368–375.
10. Zhao, J.; Chai, K.K.; Chen, Y.; Schormans, J.; Alonso-Zarate, J. Joint mode selection and resource allocation for machine-type D2D links. *Trans. Emerg. Telecommun. Technol.* **2015**, doi:10.1002/ett.3000.
11. Min, H.; Lee, J.; Park, S.; Hong, D. Capacity enhancement using an interference limited area for device-to-device uplink underlaying cellular networks. *IEEE Trans. Wirel. Commun.* **2011**, *10*, 3995–4000.
12. Yu, G.; Xu, L.; Feng, D.; Yin, R.; Li, G.Y.; Jiang, Y. Joint mode selection and resource allocation for device-to-device communications. *IEEE Trans. Commun.* **2014**, *62*, 3814–3824.
13. An, R.; Sun, J.; Zhao, S.; Shao, S. Resource allocation scheme for device-to-device communication underlying lte downlink network. In Proceedings of the 2012 International Conference on Wireless Communications & Signal Processing (WCSP), Huangshan, China, 25–27 October 2012; IEEE: Piscataway, NJ, USA, 2012; pp. 1–5.

14. Esmat, H.H.; Elmesalawy, M.M.; Ibrahim, I.I. Adaptive Resource Sharing Algorithm for Device-to-Device Communications Underlaying Cellular Networks. *IEEE Commun. Lett.* **2016**, *20*, 530–533.

15. Wang, F.; Song, L.; Han, Z.; Zhao, Q.; Wang, X. Joint scheduling and resource allocation for device-to-device underlay communication. In Proceedings of the 2013 IEEE Wireless Communications and Networking Conference (WCNC), Shanghai, China, 7–10 April 2013; IEEE: Piscataway, NJ, USA, 2013; pp. 134–139.

16. Yin, R.; Yu, G.; Zhang, H.; Zhang, Z.; Li, G.Y. Pricing-based interference coordination for D2D communications in cellular networks. *IEEE Trans. Wirel. Commun.* **2015**, *14*, 1519–1532.

17. Luo, Y.; Shi, Z.; Zhou, X.; Liu, Q.; Yi, Q. Dynamic resource allocations based on Q-learning for D2D communication in cellular networks. In Proceedings of the 2014 11th International Computer Conference on Wavelet Active Media Technology and Information Processing (ICCWAMTIP), Chengdu, China, 19–21 December 2014; IEEE: Piscataway, NJ, USA, 2014; pp. 385–388.

18. Nie, S.; Fan, Z.; Zhao, M.; Gu, X.; Zhang, L. Q-learning based power control algorithm for D2D communication. In Proceedings of the 2016 IEEE 27th Annual International Symposium on Personal, Indoor, and Mobile Radio Communications (PIMRC), Valencia, Spain, 4–8 September 2016; IEEE: Piscataway, NJ, USA, 2016; pp. 1–6.

19. Hwang, Y.; Park, J.; Sung, K.W.; Kim, S.L. On the throughput gain of device-to-device communications. *ICT Express* **2015**, *1*, 67–70.

20. Mehta, M.; Aliu, O.G.; Karandikar, A.; Imran, M.A. A self-organized resource allocation using inter-cell interference coordination (ICIC) in relay-assisted cellular networks. *arXiv* **2011**, arXiv:1105.1504.

21. Semasinghe, P.; Hossain, E.; Zhu, K. An evolutionary game for distributed resource allocation in self-organizing small cells. *IEEE Trans. Mob. Comput.* **2015**, *14*, 274–287.

22. Graziosi, F.; Santucci, F. A general correlation model for shadow fading in mobile radio systems. *IEEE Commun. Lett.* **2002**, *6*, 102–104.

23. Zulhasnine, M.; Huang, C.; Srinivasan, A. Penalty function method for peer selection over wireless mesh network. In Proceedings of the 2010 IEEE 72nd Vehicular Technology Conference Fall (VTC 2010-Fall), Ottawa, ON, Canada, 6–9 September 2010; IEEE: Piscataway, NJ, USA, 2010; pp. 1–5.

24. Khan, M.I. Resource-aware task scheduling by an adversarial bandit solver method in wireless sensor networks. *EURASIP J. Wirel. Commun. Netw.* **2016**, *2016*, doi:10.1186/s13638-015-0515-y.

25. Kaelbling, L.P.; Littman, M.L.; Moore, A.W. Reinforcement learning: A survey. *J. Artif. Intell. Res.* **1996**, *4*, 237–285.

26. Sutton, R.S.; Barto, A.G. *Reinforcement Learning: An Introduction*; MIT Press: Cambridge, MA, USA, 1998; Volume 1.

27. Khan, M.I.; Rinner, B. Performance analysis of resource-aware task scheduling methods in Wireless sensor networks. *Int. J. Distrib. Sensor Netw.* **2014**, *10*, 765182.

28. Chen, M.; Chen, J.; Ma, Y.; Yu, T.; Wu, Z. Base station assisted device-to-device communications for content update network. In Proceedings of the 2015 First International Conference on Computational Intelligence Theory, Systems and Applications (CCITSA), Yilan, Taiwan, 10–12 December 2015; IEEE: Piscataway, NJ, USA, 2015; pp. 23–28.

29. Shah, K.; Kumar, M. Distributed independent reinforcement learning (DIRL) approach to resource management in wireless sensor networks. In Proceedings of the 2007 IEEE International Conference on Mobile Adhoc and Sensor Systems (MASS), Pisa, Italy, 8–11 October 2007; IEEE: Piscataway, NJ, USA, 2007; pp. 1–9.

30. Khan, M.I.; Rinner, B. Energy-aware task scheduling in wireless sensor networks based on cooperative reinforcement learning. In Proceedings of the 2014 IEEE International Conference on Communications Workshops (ICC), Sydney, NSW, Australia, 10–14 June 2014; IEEE: Piscataway, NJ, USA, 2014; pp. 871–877.

31. Jain, R.; Chiu, D.M.; Hawe, W.R. *A Quantitative Measure of Fairness and Discrimination for Resource Allocation in Shared Computer System*; Digital Equipment Corporation: Hudson, MA, USA, 1984; Volume 38.

future internet

MDPI

Article

Quality of Service Based NOMA Group D2D Communications

Asim Anwar [†], Boon-Chong Seet *,[†] and Xue Jun Li [†]

Department of Electrical and Electronic Engineering, Auckland University of Technology, Auckland 1010, New Zealand; asim.anwar@aut.ac.nz (A.A.); xuejun.li@aut.ac.nz (X.J.L.)
* Correspondence: boon-chong.seet@aut.ac.nz
† These authors contributed equally to this work.

Received: 6 October 2017; Accepted: 26 October 2017; Published: 1 November 2017

Abstract: Non-orthogonal multiple access (NOMA) provides superior spectral efficiency and is considered as a promising multiple access scheme for fifth generation (5G) wireless systems. The spectrum efficiency can be further enhanced by enabling device-to-device (D2D) communications. In this work, we propose quality of service (QoS) based NOMA (Q-NOMA) group D2D communications in which the D2D receivers (DRs) are ordered according to their QoS requirements. We discuss two possible implementations of proposed Q-NOMA group D2D communications based on the two power allocation coefficient policies. In order to capture the key aspects of D2D communications, which are device clustering and spatial separation, we model the locations of D2D transmitters (DTs) by Gauss–Poisson process (GPP). The DRs are then considered to be clustered around DTs. Multiple DTs can exist in proximity of each other. In order to characterize the performance, we derive the Laplace transform of the interference at the probe D2D receiver and obtain a closed-form expression of its outage probability using stochastic geometry tools. The performance of proposed Q-NOMA group D2D communications is then evaluated and benchmarked against conventional paired D2D communications.

Keywords: quality of service; device-to-device communication; non-orthogonal multiple access; stochastic geometry

---

1. Introduction

With the advancement in mobile communication research, the usage of cellular technology has spread beyond voice and simple data transfer to high data rate, delay sensitive, and loss tolerant multimedia applications. Despite the fast growth of fourth generation (4G) systems, the current spectrum resources are still scarce to meet the ever increasing subscribers' demands for bandwidth and resource hungry applications, with vigorous requirements of seamless connectivity, anywhere and anytime. These trends compelled wireless researchers from the academia and industry to define new paradigm technologies and structures to achieve the goals of fifth generation (5G) systems [1,2].

In order to realize the concept of 5G into reality, many enabling technologies are proposed, among which millimetre waves, massive multiple-input multiple-output (MIMO), full-duplex (FD), heterogeneous deployments and software-defined networks have captured the attention of both academia and industry [2,3]. Nevertheless, the role of multiple access scheme always remains a vital factor in cellular networks in order to enhance the system capacity in a cost effective manner, while utilizing the bandwidth in such a way that overall spectral efficiency will be increased [4].

Non-orthogonal multiple access (NOMA) is considered as a promising multiple access candidate for future fifth generation (5G) wireless systems due to its potential of improving spectrum efficiency [5,6]. Unlike conventional orthogonal multiple access (OMA), NOMA superimposes message signals of different users in power domain and send this conglomerate signal using the same time, frequency or code resource [7,8]. Successive interference cancellation (SIC) technique is employed at each receiver to cancel the intra-user interference [9].

Apart from NOMA, another emerging technique to enhance the spectral efficiency is device-to-device (D2D) communications [10], which has the ability to improve the spectral efficiency of the conventional cellular network by sharing same spectrum resources among cellular users and D2D pairs [11–13].

By introducing aforementioned two concepts and their potential to improve spectrum efficiency, it is natural to investigate the application of NOMA scheme to D2D communications.

1.1. Related Work and Motivation

Recently, a FD D2D aided cooperative NOMA is proposed in [14]. The base station (BS) sends a NOMA signal to one NOMA-strong and one NOMA-weak user, where the strong user is equipped with full-duplex ability. By invoking D2D communication between strong and weak NOMA user pair, the authors proposed to improve the outage performance of the weak user with the aid of D2D aided direct and cooperative transmissions. However, they only considered a single-cell scenario where NOMA is conducted at BS while strong and weak users communicate via conventional paired D2D communication. The authors in [15] developed an analytical framework based on stochastic geometry to analyse cellular networks with underlay D2D communications. The D2D users are also equipped with FD transceivers and can operate in FD mode. The authors proposed criteria to select between FD and D2D modes of operation. They derived closed-form expression for outage probability to evaluate the performance of cellular and D2D users. However, they modeled spatial topology of D2D users by a Poisson point process (PPP), which may not be realistic distribution choice for D2D users. The reason is that PPP cannot capture the features of device clustering and spatial separation of D2D communications due to its completely random nature [16]. Furthermore, NOMA protocol is not applied for both cellular and D2D communications. A relay assisted diversity communication is proposed in [17]. The proposed analytical framework analyses the frame error probability performance by considering the effects of node locations, link characteristics, power allocation, diversity methods and distributed coding and constellation signaling. However, they considered an OMA based communication between single source and destination assisted by three relay nodes. The authors in [18] considered a downlink multiuser MIMO NOMA celullar network with underlaid D2D communications. They proposed two beamforming schemes in order to eliminate the inter-beam interference and the one caused to D2D users by BS transmission. In addition, they formulated an optimization problem to jointly study the performance of both cellular and D2D users. A potential limitation to their approach is that the considered system model is limited to single cell and no specific random distribution is utilised to model the spatial topologies of cellular and D2D users. Hence, it is not straightforward to generalize the results for a case of multi-cell network. Furthermore, they considered a paired D2D communication where D2D users do not apply NOMA protocol to communicate with each other.

In [19], the authors considered a NOMA-based D2D communications and introduced the concept of D2D group, where a D2D transmitter (DT) is communicating with multiple D2D receivers (DRs) using NOMA protocol. The authors proposed an optimal resource allocation strategy for interference management that enables to realize the NOMA-based D2D group communications. Although the concept of NOMA group D2D was introduced in [19], their system model was comprised of single-cell and lack of interference characterization at DR. Furthermore, it requires interference modeling and performance evaluation at the DR in order to extend the concept of NOMA group D2D to a general scenario, where DTs and DRs are distributed in the entire network. The authors in [20] further proposed cooperative hybrid automatic repeat request (HARQ) assisted NOMA in large-scale D2D networks. They studied the outage and throughput performance of the D2D users under the considered network setting and demonstrated that cooperative HARQ assisted NOMA achieves lower outage probability than non-cooperative case and OMA scheme. However, their NOMA based D2D network model is restricted to the two-user case only i.e., they only considered two-user

NOMA transmission from D2D sources. Furthermore, they considered a significant difference between channels of two D2D users by assuming that one user is always closer to D2D source compared to the other user. This assumption may not always hold, particularly in the scenario of NOMA based D2D communications because DRs are clustered around DTs and are located in proximity of each other and hence DRs may have very similar channel conditions.

In the light of the abovementioned discussion, very little attention was paid to investigate NOMA group D2D communications. Hence, the aforementioned gaps and shortcomings motivated us to investigate and analyse the performance of NOMA based group D2D communications under interference limited scenarios. In order to capture the key features of D2D communications, i.e., device clustering and spatial separation, the DTs are considered to be randomly distributed over $\mathbb{R}^2$ according to Gauss–Poisson process (GPP), while the DRs are assumed to be randomly clustered around DTs. The reason for choosing GPP is that it is a relatively simpler cluster point process that maintains good trade-off between modeling accuracy and analytical tractability. Therefore, in the context of D2D communications, GPP provides more realistic modeling approach against PPP case by capturing the clustering behavior of D2D devices [21].

Furthermore, the current approach to order users in NOMA group D2D communications is based on the channel gains of the DRs. This ordering approach may not be suitable to D2D communication scenario under which the DRs in the same group are clustered around a common DT and are located in proximity of each other. Hence, the channel conditions of the DRs located in the same D2D group would be very similar. Consequently, this ordering strategy may result in very similar power allocation, which could limit the gains of applying NOMA to D2D communications [7]. In the context of NOMA, there are few works that use quality of service (QoS) based ordering [22–24]. They mainly focused on cellular networks and are limited to the two-user case only. In this paper, we propose and analyse QoS based NOMA (Q-NOMA) group D2D communications and make an attempt to fill the aforementioned gaps in literature. To the best of our knowledge, it is the first time that Q-NOMA is proposed and analysed to realize group D2D communications under interference limited scenario.

1.2. Contributions

The main contributions of this work are as follows:

- We propose Q-NOMA group D2D communications in which D2D users are randomly distributed over the entire two-dimensional plane. Unlike the existing proposals, we order the DRs according to their QoS requirements, which is more appropriate for the D2D communications scenario. Furthermore, in contrast to PPP, which is most commonly used to model D2D users (both DTs and DRs), we model the spatial topology of DTs by GPP and DRs are considered to be randomly clustered around DTs. These spatial distributions of DTs and DRs are suitable to analyse the proposed network with any number of D2D users. In addition, based on the QoS ordering, we propose two policies to compute power allocation coefficients that could lead to two implementations of the proposed Q-NOMA group D2D communications.
- We derive the interference distribution at the probe DR by utilizing the results from stochastic geometry. The Laplace transform of interference over GPP is derived in [21], which involves complex double integrals. In order to obtain useful insights, a major step in characterizing the interference is the approximation of integrals in the interference Laplace transform by applying Gaussian–Chebyshev and Gauss–Laguerre quadratures. This approximation results in an interference Laplace transform expression, which is easy to implement.
- Based on the interference approximation results, we further derive the closed-form expression for outage probability of the DRs in the proposed Q-NOMA group D2D communications.
- We present numerical results to validate the accuracy of the derived outage results and compare the performance of the proposed Q-NOMA group D2D with conventional paired D2D communications using OMA.

1.3. Mathematical Preliminaries on Gauss–Poisson Process

The generalized GPP [21] is defined as the Poisson cluster process with homogeneous independent clustering. Denote λ as the parent process intensity. Then, each cluster in GPP can be classified as single-point or two-point cluster. Let $1 - a$ and a denote the probabilities that a cluster is single-point, and two-point, respectively. When the cluster is single-point, the point is located at the position of the parent. When the cluster is two-point, one point is located at the position of the parent while the other is randomly distributed around the parent with some probability density function (PDF) $f_u(\cdot)$, where u is the inter-point distance in two-point cluster.

2. System Model

Consider inband D2D communications with an overlay cellular network. We consider a frequency reuse factor of one among D2D users to achieve better spectrum efficiency. With this setting, every D2D transmission by a DT is subjected to interference from other active DTs. We consider a composite fading and path loss channel model between every DT and DR. In this work, we assume that the power fading coefficients are independent and identically distributed (i.i.d) with exponential distribution of unit mean, and adopt a path loss model of $d^{-\alpha}$, where d is the distance between the probe DR and test DT, and α is the path loss exponent.

2.1. Spatial Distribution of D2D Users

Consider that D2D users are randomly distributed over $\mathbb{R}^2$. At any time realization, the D2D users are classified as transmitters or receivers. We consider the group D2D scenario, where each DT is communicating with multiple DRs via a NOMA scheme. We refer to a DT communicating to multiple D2D devices as a group transmitter (GT). Any D2D user can take a role of GT. We assume that the selection of GTs is performed by a BS, and it can select multiple GTs in a given cell to improve overall system capacity. We allow multiple GTs to exist in proximity of each other, where each GT is communicating to its own group of receivers. Hence, at any time realization, each selected GT forms a group/cluster containing DRs. In order to capture both inband and device clustering, we model the spatial topology of the GTs by a stationary and isotropic GPP defined on $\mathbb{R}^2$, denoted by Φ_{GT} with parent process intensity λ_{GT}. Furthermore, we model the coverage of each GT by a disc D with radius R_D. We consider that the DRs are clustered around each GT and are uniformly distributed inside coverage of GTs. An illustration of NOMA group D2D communication is presented in Figure 1. For a quick reference, a list of commonly used abbreviations is given in Table 1.

Table 1. List of commonly used abbreviations.

D2D	Device-to-device
DR	D2D receiver
DT	D2D transmitter
F-NOMA	Fixed NOMA
GPP	Gauss–Poisson process
GT	Group transmitter
NOMA	Non-orthogonal multiple access
PPP	Poisson point process
QoS	Quality of service
SIC	Successive interference cancellation
SNR	Signal to noise ratio

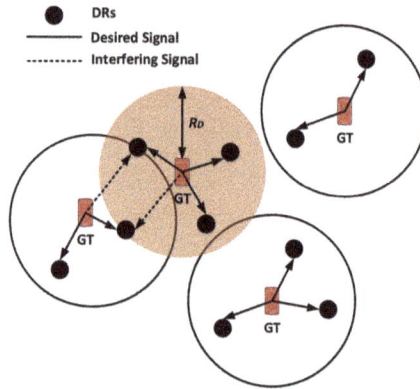

Figure 1. Example of inband non-orthogonal multiple access (NOMA) group device-to-device (D2D) communications with overlay cellular network.

2.2. Q-NOMA Group D2D Communication

Since DRs are clustered around GTs and are located in proximity of each other, therefore, the DRs connected to same GT would have very similar channel conditions. Consequently, in the context of D2D communications, ordering DRs according to their channel conditions to apply NOMA at GT may not achieve the desired multiplexing gains and fairness among DRs. Hence, in this work, we propose ordering the DRs of the GT according to QoS requirements, which are determined by their targeted data rates.

Let there be a total of M DRs distributed inside coverage of a test GT. The probe DR is assumed to be located at the origin, with a desired test GT at $x_0 = (d, 0)$ with $d \neq 0$. Without loss of generality, we assume that the DRs are ordered as $R_1 \leq \dots \leq R_M$, where R_i is the targeted data rate of DR i, $1 \leq i \leq M$. Correspondingly, the power allocation coefficients are sorted as $\beta_1 \geq \dots \geq \beta_M$.

The aforementioned procedure to order DRs of the GT and compute their power allocation coefficients according to users' targeted rates is termed as "Q-NOMA". When it is applied to D2D communications, we refer to the communication as "Q-NOMA group D2D communication".

Consider that the NOMA DR m is the probe receiver, then, the received signal from the test transmitter at the probe DR is given as:

$$y_m = h_m \sum_{i=1}^{M} \sqrt{\beta_i P_{\mathrm{GT}}} s_i + n_m,$$ (1)

where P_{GT} is the transmission power of test GT, s_i is the message signal of DR i and n_m is the additive white Gaussian noise (AWGN) with zero mean and variance σ^2.

2.3. Power Allocation Coefficients Policies

The optimal power and resource allocation improve overall performance and utilise the system resources efficiently. However, the optimum power allocation strategies proposed in existing literature, for example, [25], cannot be directly applied to the current work because of significant difference in system model or underlying transmission method. Therefore, in this sub-section, we discuss two simple methods to compute power allocation coefficients $\{\beta_i\}_{i=1}^{M}$ that would lead to two possible implementations of Q-NOMA group D2D communications.

2.3.1. Policy I

The Policy I utilises targeted rates of DRs to compute the power allocation coefficients. Similar to [7], the power allocation coefficient for DR i under Policy I is computed as:

$$\beta_i = \frac{1/R_i}{\sum_{j=1}^{M} \frac{1}{R_j}}. \tag{2}$$

The intuition behind Equation (2) is that power allocation coefficients could be utilised to maintain fairness among DRs [3,26]. In case they are computed in proportion with users' targeted rates, then the highest ordered DR M would result in the highest SIC decoding order with maximum signal-to-interference-plus-noise ratio (SINR) threshold among all ordered DRs. Consequently, all of the lower ordered DRs always require the maximum SINR threshold dictated by the user M, which would result in a biased treatment of lower ordered users (with lower targeted rates). Hence, in order to avoid this biasness, we propose to compute the power allocation coefficients as given by Equation (2).

2.3.2. Policy II

The concept of fixed NOMA (F-NOMA) was proposed in [27], where the power allocation coefficients are fixed and are computed based on the given user ordering i.e., it does not utilise the actual channel gains to compute $\{\beta_i\}_{i=1}^{M}$. Similar to [27], we adopt a Policy II to compute $\{\beta_i\}_{i=1}^{M}$ for conventional NOMA that does not utilise the actual targeted rates of DRs. The power allocation coefficient for DR i under Policy II is computed as:

$$\beta_m^F = \frac{M - m + 1}{\mu}, \tag{3}$$

where β_m^F represents the power allocation coefficient of DR m in Policy II and μ is selected in such a way that $\sum_{i=1}^{M} \beta_i^F = 1$.

2.4. Interference Distribution

The reception at the probe DR from the test GT is interfered by the other GTs. The interference at probe DR is given as, $I = \sum_{x \in \Phi_{GT} \setminus x_0} |g_x|^2 d_x^{-\alpha}$, where g_x and d_x represent the Rayleigh fading channel gain and distance between probe DR and interferer at x_0, respectively. The following lemma provides the Laplace transform of the interference at probe DR.

Lemma 1. *Consider a GPP Φ_{GT} with parent process intensity λ_{GT} modeling spatial topology of the GTs in a Q-NOMA group D2D communications. Then, the Laplace transform of the interference at the probe DR conditioned at the location of test GT is given by:*

$$\mathcal{L}_I(s) = e^{-2\pi\lambda_{GT}\sum_{p=1}^{P} \Omega_p \frac{a(1-X_1(r_p))+sr_p^{-\alpha}}{1+sr_p^{-\alpha}}} \cdot \Lambda_2(d), \tag{4}$$

where $\Omega_p = \omega_p e^{r_p}$, $\omega_p = \frac{\Gamma(P+2)r_p}{P!(P+1)^2(L_{P+1}(r_p))^2}$, $L_P(\cdot)$ is the Laguerre polynomial of degree P, r_p are the roots of $L_P(\cdot)$, $X_1(\cdot)$ and $\Lambda_2(\cdot)$ are given in Equations (A7) and (A10), respectively.

Proof. See Appendix A. □

3. Outage Probability Analysis

In this section, we focus on the outage probability for the DRs in the considered Q-NOMA group D2D communication. Let τ_m and R_m denote the SINR threshold and targeted rate of DR m, respectively, where $\tau_m = 2^{R_m} - 1$. Since each DR employs SIC, the outage at DR m occurs if it does not meet the targeted rate of any higher order DR j, where $1 \le j < m$. Denote $\zeta_{m \to j} = \{\tilde{R}_{m,j} < R_j\}$ as the outage event at DR m due to decoding of DR j, where $\tilde{R}_{m,j}$ is the achievable rate of user j at DR m. The outage event $\zeta_{m \to j}$ can be expressed as:

$$
\begin{aligned}
\zeta_{m \to j} &= \{\tilde{R}_{m,j} < R_j\} \\
&= \left\{ \log_2 \left(1 + \frac{h_m \beta_j P}{h_m P \sum_{i=j+1}^{M} \beta_i + I + \sigma^2} \right) < R_j \right\} \\
&= \left\{ h_m < \frac{\varphi_j(\rho I + 1)}{\rho_t} \right\},
\end{aligned}
\tag{5}
$$

where $\varphi_j = \frac{\tau_j}{\beta_j - \tau_j \sum_{i=j+1}^{M} \beta_i}$, $\rho = \frac{P_I}{\sigma^2}$, $\rho_t = \frac{P_{GT}}{\sigma^2}$ is the transmit signal-to-noise ratio (SNR) and P_I is the maximum received interference power at the probe DR.

Next, define $\varphi_m^{\max} = \max\{\varphi_1, ..., \varphi_m\}$. Based on Equation (3), the outage probability at the DR m can be given as:

$$
\begin{aligned}
\mathbb{P}_m &= \Pr\left(h_m < \frac{\varphi_m^{\max}(\rho I + 1)}{\rho_t} \right) \\
&= \mathbb{E}_I\left[F_{h_m}\left(\frac{\varphi_m^{\max}(\rho x + 1)}{\rho_t} \right) \right],
\end{aligned}
\tag{6}
$$

where F_{h_m} is the cumulative distribution function (CDF) of h_m.

Note that the set $\{h_i\}, i = 1, ..., M$, of channel gains is not ordered because the users are sorted in ascending order of their targeted rates. Since the channel gains are i.i.d. random variables with common CDF F_h, Equation (4) can be re-written as:

$$
\mathbb{P}_m = \mathbb{E}_I\left[F_h\left(\frac{\varphi_m^{\max}(\rho x + 1)}{\rho_t} \right) \right].
\tag{7}
$$

Consequently, the outage probability of DR m is provided in the following theorem.

Theorem 1. *The outage probability of DR m in the Q-NOMA group D2D communications is derived as:*

$$
\mathbb{P}_m = \sum_{l=1}^{L} b_l e^{-\frac{\varphi_m^{\max} c_l}{\rho_t}} \mathcal{L}_I\left(\frac{\varphi_m^{\max} c_l \rho}{\rho_t} \right),
\tag{8}
$$

where $b_l = \omega_l \sqrt{1 - \phi_l^2}(1 + \phi_l)$, $\omega_l = \frac{\pi}{L}$, $c_l = \left(\frac{R_D}{2}(1 + \phi_l) \right)^\alpha$, $\phi_l = \cos\left(\frac{(2l-1)\pi}{2L} \right)$ *and L is the complexity-accuracy trade-off parameter.*

Proof. See Appendix B. □

Note that, due to the presence of interferers in the network, similar to [20,28], the derived outage probability in Equation (8) is a function of variables $\varphi_m^{\max}$ and interference Laplace transform $\mathcal{L}_I$. This is different from existing works that analyse NOMA wireless systems under no interference where outage probability is mainly a function of $\varphi_m^{\max}$. For reference, please see [14,27,29]. Furthermore, it is worthy to note that the current outage analysis approach remains valid if more complex cluster models (Poisson cluster process, etc.) are adopted for modeling spatial distributions of DTs and DRs. In that case, the $\mathcal{L}_I$ term will be replaced by the Laplace transform of the interference for the adopted model. Intuitively, $\mathcal{L}_I$ is performing a form of scaling in Equation (8) and hence the conclusions are expected to remain the same if the spatial distribution model(s) of the DTs and DRs is changed. However, the exact impact on outage probability when more complex cluster processes model is adopted for modeling DT and DR locations would require further study, and we plan to investigate them in our future work.

4. Numerical Results and Discussion

This section presents the numerical results to evaluate the performance of the considered network as well as to validate the accuracy of the derived expression in Equation (8) of Section 3. As shown in Table 2, simulation parameters used are similar to those in [29], unless otherwise stated. Furthermore, we adopt Policy I as a default policy to compute power allocation coefficients, unless otherwise stated.

Table 2. Simulation parameters.

Parameter	Description	Value
M	Total users	3
$\{R_m\}_{m=1}^{M}$	Users' targeted rates	$\{0.7, 1.1, 2\}$
R_D	Coverage of GT	10 m
α	Path loss exponent	4
λ_{GT}	Intensity of GTs	10^{-4}
ρ_t	SNR range	(5–40) dB
L, N, V, Q, S	Gaussian-Chebyshev parameters	5
P	Degree of Gauss-Laguerre polynomial	5
d	Distance between probe DR and GT	5 m

4.1. Impact of R_D on Outage Probability

Figure 2 presents the impact of varying coverage radius R_D of test GT on the outage probability of ordered DRs as a function of SNR. The derived outage results in Equation (8) are shown to be in good agreement with the Monte Carlo simulations.

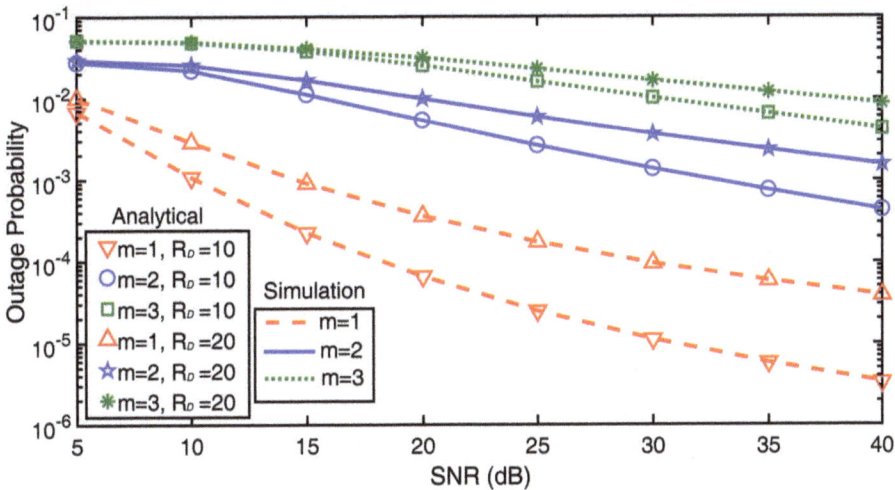

Figure 2. Impact of R_D on outage probability.

Several observations can be made from the results in Figure 2: (1) increasing the coverage radius of GT results in a higher outage probability because of a larger path loss; (2) different ordered users have distinct decreasing slopes of outage probability because of different targeted rates; (3) the higher order DRs in our proposed Q-NOMA group D2D communications achieve better outage performance because they have smaller targeted rates. This is different from conventional NOMA that orders users based on channel conditions, and where the higher ordered users have larger outage probabilities due to poor channel conditions [27,28].

4.2. Impact of d on Outage Probability

The impact of varying distance between probe DR and test GT on the outage performance is investigated in Figure 3. It can be observed that varying d has a larger impact on the average achievable outage probability at lower SNR value of 5 dB. This is intuitively plausible because increasing transmission power results in improved SINR at the receiver and hence better outage performance.

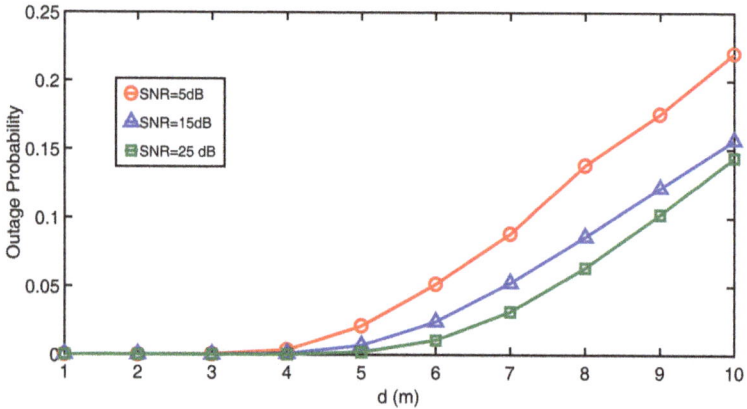

Figure 3. Impact of d on outage probability.

4.3. Comparison between Paired and Grouped D2D Communications

The average outage probability achieved by Q-NOMA group D2D communications under different path loss exponents is shown in Figure 4. The performance of paired D2D communication based on OMA is also presented in the figure as a benchmark for comparison. It can be observed that Q-NOMA group D2D achieves overall lower outage probability than the paired D2D communication for different values of path loss. This is because, as opposed to paired D2D, Q-NOMA group D2D communication uses only single transmission, which results in better SINR at the DRs under an interference limited scenario.

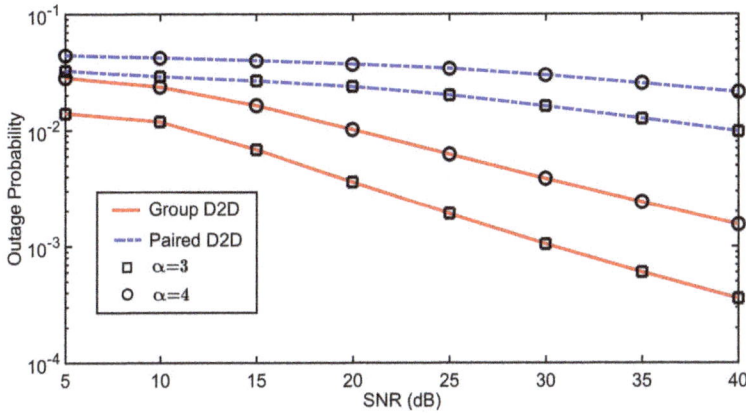

Figure 4. Outage comparison between paired and group D2D.

4.4. Comparison between Two Implementations of Q-NOMA Group D2D Communications

In this section, we compare the performance of two possible implementations of Q-NOMA group D2D communications based on power allocation coefficients Policies I and II. The results for Q-NOMA group D2D communication under Policy I are also obtained by utilising Equation (8), with the exception that $\{\varphi_m^{max}\}_{m=1}^M$ are computed by using fixed power allocation coefficients $\{\beta_m^F\}_{m=1}^M$. We consider five different cases for users' targeted rates and compared the outage probabilities achieved by the Q-NOMA group D2D communications under Policies I and II at an SNR of 25 dB. Table 3 summarizes the outage comparison between these two implementations of the group D2D communications. The power allocation coefficients for Policy I are derived as $\beta_m^F = \{0.5, 0.33, 0.17\}_{m=1}^M$, whereas those for Policy II are calculated using Equation (2) based on users' targeted rates and are shown in Table 3.

Table 3. Average outage comparison between two implementations of quality of service based non-orthogonal multiple access (Q-NOMA) group device-to-device (D2D) communications.

Case	Targeted Rates, $\{R_m\}_{m=1}^3$	Power Allocation Coefficients, $\{\beta_m\}_{m=1}^3$	Policy I	Policy II
1.	$\{1, 2.5, 3\}$	$\{0.58, 0.24, 0.18\}$	0.04	1
2.	$\{1, 1.5, 2\}$	$\{0.47, 0.3, 0.23\}$	0.002	0.04
3.	$\{0.5, 1.5, 2\}$	$\{0.63, 0.21, 0.16\}$	1	1
4.	$\{0.9, 1, 2\}$	$\{0.42, 0.38, 0.2\}$	1	0.001
5.	$\{0.2, 2, 2.2\}$	$\{0.85, 0.08, 0.07\}$	0.006	0.002

It is observed from the results in Table 3 that Q-NOMA under Policy I achieves lower outage probability in cases 1 and 2, whereas it obtains better outage performance in cases 4 and 5 for Policy II. In addition, the results in case 3 indicate the importance of proper power and rate allocation to avoid situations of complete outage. Moreover, it can be observed that the Q-NOMA group D2D communication under Policy I performs better in cases where the users' targeted rates are significantly apart. This is more suitable to D2D communication scenarios where different users may have diverse QoS requirements.

In order to extend the analysis beyond the case of $M > 3$ users, we further present in Table 4 the average outage comparison between paired D2D and Q-NOMA group D2D communications (based on power allocation coefficients Policies I and II). We consider $M = 5, 7, 9$ with $R_m \sim \mathcal{U}(0.1, 2), \forall m \in M$ and $\rho_t = 25$ dB, where $\mathcal{U}(\cdot, \cdot)$ represents the random uniform distribution function. It can be observed that the results in Table 4 are consistent with those presented in Figure 4, for example, Q-NOMA group D2D communications consistently achieve lower outage probability than paired D2D communications.

Table 4. Average outage comparison between paired device-to-device (D2D) and quality of service based non-orthogonal multiple access (Q-NOMA) group D2D communications ($M > 3$).

Total Users, M	Q-NOMA Policy I	Q-NOMA Policy II	Paired D2D
5	0.003	0.07	0.08
7	0.007	0.06	0.12
9	0.01	0.1	0.3

5. Conclusions

In this paper, we have proposed Q-NOMA group D2D communications. In order to study the performance of the proposed network, we first derive the Laplace transform of the interference expression, based on which we further derive the closed-form expression for outage probability to analyse the performance of the DRs in the proposed Q-NOMA group D2D communications. The results show that the proposed Q-NOMA group D2D achieves overall lower outage probability than its counterpart paired D2D communication. Furthermore, based on two power allocation coefficient policies, we have presented the comparison between two possible implementations of the proposed Q-NOMA group D2D communications. Due to the similar channel conditions and diverse QoS requirements of DRs, the results show that the proposed Q-NOMA implementation based on Policy I is more realistic and suitable than one based on Policy II for group D2D communications. As future work, we plan to investigate and analyse the performance of proposed Q-NOMA group D2D communications in underlay cellular mode, where additional interferences from base stations and cellular users will be considered. Another interesting future direction is to extend the current network model to a framework of MIMO systems. This would require a robust precoder and MIMO-NOMA design to improve the system performance and capacity while mitigating the inter and intra beam interference efficiently.

Acknowledgments: This work has been supported by a PhD Fee and Stipend Scholarship for Asim Anwar from the School of Engineering, Computer and Mathematical Sciences at Auckland University of Technology.

Author Contributions: Asim Anwar and Boon-Chong Seet conceived and designed the study. Asim Anwar derived the Lemmas and Theorems, conducted the simulations and drafted the paper. Boon-Chong Seet and Xue Jun Li critically reviewed the results, proofread and revised the paper.

Conflicts of Interest: The authors declare no conflict of interest.

Appendix A. Proof of Lemma 1

In the case of GPP, when desired transmitter is at $x_0 \in \Phi_{GT}$, the $\mathcal{L}_{\mathcal{I}}(s)$ is given by Equation (34) of [21]:

$$\mathcal{L}_{\mathcal{I}}(s) = \Lambda_1 \cdot \Lambda_2, \tag{A1}$$

where

$$\Lambda_1 = \exp\left\{ 2\pi\lambda_{GT} \int_0^\infty \left[\frac{1-a}{1+sr^{-\hat{\alpha}}} + \frac{a}{1+sr^{-\hat{\alpha}}} \cdot \int_0^\infty \int_0^{2\pi} \frac{\tau f_u(\tau)d\psi}{1+s(r^2+\tau^2+2r\tau\cos(\psi))^{-\alpha/2}}d\tau - 1 \right] r dr \right\}, \tag{A2}$$

and

$$\Lambda_2 = \frac{1-a}{1+a} + \frac{2a}{1+a} \int_0^\infty \int_0^{2\pi} \frac{\tau f_u(\tau)d\psi}{1+s\left(d^2+\tau^2+2d\tau\cos(\psi)\right)^{-\alpha/2}}d\tau, \tag{A3}$$

where $1 - a$ and a are the probabilities of having one and two transmitters in a group, respectively.

Now let us take X_1 in Equation (A2) as:

$$X_1 = \int_0^\infty \int_0^{2\pi} \frac{\tau f_u(\tau) d\psi}{1 + s\left(r^2 + \tau^2 + 2r\tau \cos(\psi)\right)^{-\alpha/2}} d\tau$$

$$= \frac{2}{R_D^2} \int_0^{R_D} \tau^2 \underbrace{\int_0^{2\pi} \frac{d\psi}{1 + s\left(r^2 + \tau^2 + 2r\tau \cos(\psi)\right)^{-\alpha/2}}}_{X_2} d\tau, \tag{A4}$$

where $f_u(\tau) = \frac{2\tau}{R_D^2}$ if $0 \le \tau \le R_D$.

It is challenging to solve integral X_2 in Equation (A4). As such, we approximate it by applying the Gaussian–Chebyshev quadrature as [30]:

$$X_2 \approx \sum_{n=1}^N \frac{\varphi_n}{1 + s\left(r^2 + \tau^2 + 2r\tau \cos(\pi t_n)\right)^{-\alpha/2}}, \tag{A5}$$

where $\varphi_n = \pi \omega_n \sqrt{1 - \theta_n^2}$, $\omega_n = \frac{\pi}{N}$, $\theta_n = \cos\left(\frac{2n-1}{2N}\pi\right)$, $t_n = 1 + \theta_n$ and N is the complexity-accuracy tradeoff parameter.

Based on Equation (A5), X_1 can now be expressed as:

$$X_1 = \frac{2}{R^2} \int_0^{R_D} \sum_{n=1}^N \frac{\varphi_n \tau^2}{1 + s\left(r^2 + \tau^2 + 2r\tau \cos(\pi t_n)\right)^{-\alpha/2}} d\tau. \tag{A6}$$

Note that it is challenging to solve Equation (A6) analytically. In order to obtain insightful results, we approximate it by applying Gaussian–Chebyshev quadrature as:

$$X_1(r) \approx \sum_{v=1}^V \sum_{n=1}^N \frac{2R_D \varphi_n \omega_v \sqrt{1 - \vartheta_v^2} k_v^2}{1 + s\left(r^2 + R_D^2 k_v^2 + 2r R_D k_v^2 \cos(\pi t_n)\right)^{-\alpha/2}}, \tag{A7}$$

where $\omega_v = \frac{\pi}{V}$, $\vartheta_v = \cos\left(\frac{2v-1}{2V}\pi\right)$, $k_v = \frac{1}{2}(\vartheta_v + 1)$ and V is the complexity-accuracy tradeoff parameter.

Based on Equation (A7), Λ_1 in Equation (A2) is re-written as:

$$\Lambda_1 = \exp\left\{2\pi\lambda_{GT} \int_0^\infty \left[\frac{1-a}{1 + sr^{-\alpha}} + \frac{aX_1(r)}{1 + sr^{-\alpha}} - 1\right] rdr\right\}$$

$$= \exp\left\{-2\pi\lambda_{GT} \int_0^\infty \frac{a\left(1 - X_1(r)\right) + sr^{-\alpha}}{1 + sr^{-\alpha}}\right\}. \tag{A8}$$

Next, we apply Gauss–Laguerre quadrature to approximate the integral in Equation (A8). Hence, Λ_1 can be expressed after approximation as:

$$\Lambda_1 = e^{-2\pi\lambda_{GT} \sum_{p=1}^P \Omega_p \frac{a\left(1 - X_1(rp)\right) + sr_p^{-\alpha}}{1 + sr_p^{-\alpha}}}. \tag{A9}$$

Following the same approximation procedure for X_1 and applying Gaussian–Chebyshev quadrature twice, Λ_2 in Equation (A3) is given as:

$$\Lambda_2(d) = \frac{1-a}{1+a} + \frac{2a}{1+a} \sum_{j=1}^S \sum_{i=1}^Q \frac{\varrho_i \xi_j}{1 + s\left(d^2 + R_D^2 z_j^2 + 2d R_D z_j \cos(\pi x_i)\right)^{-\alpha/2}}, \tag{A10}$$

where $\varrho_i = \pi \omega_i \sqrt{1 - \eta_i^2}$, $\omega_i = \frac{\pi}{Q}$, $\eta_i = \cos\left(\frac{2i-1}{2Q}\pi\right)$, $x_i = \eta_i + 1$, $\xi_j = 2R_D \omega_j \sqrt{1 - \Theta_j^2 z_j^2}$, $\Theta_j = \cos\left(\frac{2j-1}{2S}\pi\right)$, $z_j = \frac{1}{2}(\Theta_j + 1)$, $\omega_j = \frac{\pi}{S}$ and Q, S are the complexity-accuracy trade-off parameters.

Finally, the result in Lemma1 is obtained by multiplying Equations (A9) and (A10). $\qquad\square$

Appendix B. Proof of Theorem 1

In order to obtain $\mathbb{P}_m$, we require F_h. Since, all wireless links exhibit Rayleigh fading and the DRs are uniformly distributed inside disc D centered at the location of test GT, the CDF F_h can be expressed as [31]:

$$F_h(y) = \frac{2}{R_D^2} \int_0^{R_D} \left(1 - e^{-z^\alpha y}\right) z\,dz. \tag{A11}$$

It is challenging to solve the above integral. As such, we approximate it by applying Gaussian-Chebyshev quadrature as:

$$F_h(y) = \sum_{l=1}^{L} b_l e^{-c_l y}. \tag{A12}$$

Based on Equation (A12), $\mathbb{P}_m$ in Equation (5) can be expressed as:

$$
\begin{aligned}
\mathbb{P}_m &= \int_0^\infty F_h\left(\frac{\varphi_m^{\max}(\rho x + 1)}{\rho_t}\right) f_I(x)\,dx \\
&= \sum_{l=1}^{L} \int_0^\infty e^{-\frac{c_l \varphi_m^{\max}(\rho x + 1)}{\rho_t}} f_I(x)\,dx \\
&= \sum_{l=1}^{L} b_l e^{-\frac{c_l \varphi_m^{\max}}{\rho_t}} \int_0^\infty e^{-\frac{c_l \varphi_m^{\max} \rho x}{\rho_t}}\,dx \\
&= \sum_{l=1}^{L} b_l e^{-\frac{c_l \varphi_m^{\max}}{\rho_t}} \mathcal{L}_I\left(\frac{c_l \varphi_m^{\max} \rho}{\rho_t}\right),
\end{aligned}
\tag{A13}
$$

where $f_I(x)$ is the PDF of interference I and the last step follows from the definition of Laplace transform. This proves the result in Theorem 1. $\qquad\square$

References

1. Akyildiz, I.F.; Nie, S.; Lin, S.C.; Chandrasekaran, M. 5G roadmap: 10 key enabling technologies. *Comput. Netw.* **2016**, *106*, 17–48.
2. Wong, V.W.S.; Schober, R.; Ng, D.W.K.; Wang, L.C. *Key Technologies for 5G Wireless Systems*; Cambridge University Press: Cambridge, UK, 2017.
3. Timotheou, S.; Krikidis, I. Fairness for Non-Orthogonal Multiple Access in 5G Systems. *IEEE Signal Process. Lett.* **2015**, *22*, 1647–1651.
4. Anwar, A.; Seet, B.C.; Li, X.J. PIC-based receiver structure for 5G downlink NOMA. In Proceedings of the 10th International Conference on Information, Communications and Signal Processing (ICICS), Singapore, 2–4 December 2015.
5. Ding, Z.; Liu, Y.; Choi, J.; Sun, Q.; Elkashlan, M.; Chih-Lin, I.; Poor, H.V. Application of Non-Orthogonal Multiple Access in LTE and 5G Networks. *IEEE Commun. Mag.* **2017**, *55*, 185–191.
6. Wei, Z.; Yuan, J.; Ng, D.W.K.; Elkashlan, M.; Ding, Z. A Survey of Downlink Non-orthogonal Multiple Access for 5G Wireless Communication Networks. *arXiv* **2016**, arXiv:abs/1609.01856.
7. Saito, Y.; Benjebbour, A.; Kishiyama, Y.; Nakamura, T. System-level performance evaluation of downlink non-orthogonal multiple access (NOMA). In Proceedings of the IEEE 24th Annual International Symposium on Personal, Indoor, and Mobile Radio Communications (PIMRC), London, UK, 8–11 September 2013.
8. Lv, L.; Chen, J.; Ni, Q.; Ding, Z. Design of Cooperative Non-Orthogonal Multicast Cognitive Multiple Access for 5G Systems: User Scheduling and Performance Analysis. *IEEE Trans. Commun.* **2017**, *65*, 2641–2656.

9. Kimy, B.; Lim, S.; Kim, H.; Suh, S.; Kwun, J.; Choi, S.; Lee, C.; Lee, S.; Hong, D. Non-orthogonal Multiple Access in a Downlink Multiuser Beamforming System. In Proceedings of the IEEE Military Communications Conference (MILCOM), San Diego, CA, USA, 18–20 November 2013.

10. Andrews, J.G.; Buzzi, S.; Choi, W.; Hanly, S.V.; Lozano, A.; Soong, A.C.K.; Zhang, J.C. What Will 5G Be? *IEEE J. Sel. Areas Commun.* **2014**, *32*, 1065–1082.

11. Fodor, G.; Dahlman, E.; Mildh, G.; Parkvall, S.; Reider, N.; Miklós, G.; Turányi, Z. Design aspects of network assisted device-to-device communications. *IEEE Commun. Mag.* **2012**, *50*, 170–177.

12. Doppler, K.; Rinne, M.; Wijting, C.; Ribeiro, C.B.; Hugl, K. Device-to-device communication as an underlay to LTE-advanced networks. *IEEE Commun. Mag.* **2009**, *47*, 42–49.

13. Lin, X.; Andrews, J.G.; Ghosh, A. Spectrum Sharing for Device-to-Device Communication in Cellular Networks. *IEEE Trans. Wirel. Commun.* **2014**, *13*, 6727–6740.

14. Zhang, Z.; Ma, Z.; Xiao, M.; Ding, Z.; Fan, P. Full-Duplex Device-to-Device-Aided Cooperative Nonorthogonal Multiple Access. *IEEE Trans. Veh. Technol.* **2017**, *66*, 4467–4471.

15. Ali, K.S.; ElSawy, H.; Alouini, M.S. Modeling Cellular Networks With Full-Duplex D2D Communication: A Stochastic Geometry Approach. *IEEE Trans. Commun.* **2016**, *64*, 4409–4424.

16. Afshang, M.; Dhillon, H.S. Spatial modeling of device-to-device networks: Poisson cluster process meets Poisson Hole Process. In Proceedings of the 49th Asilomar Conference on Signals, Systems and Computers, Pacific Grove, CA, USA, 8–11 November 2015.

17. Palombara, C.L.; Tralli, V.; Masini, B.M.; Conti, A. Relay-Assisted Diversity Communications. *IEEE Trans. Veh. Technol.* **2013**, *62*, 415–421.

18. Sun, H.; Xu, Y.; Hu, R.Q. A NOMA and MU-MIMO Supported Cellular Network with Underlaid D2D Communications. In Proceedings of the IEEE 83rd Vehicular Technology Conference (VTC Spring), Nanjing, China, 15–18 May 2016.

19. Zhao, J.; Liu, Y.; Chai, K.K.; Chen, Y.; Elkashlan, M.; Alonso-Zarate, J. NOMA-Based D2D Communications: Towards 5G. In Proceedings of the 2016 IEEE Global Communications Conference (GLOBECOM), Washington, DC, USA, 4–8 December 2016.

20. Shi, Z.; Ma, S.; ElSawy, H.; Yang, G.; Alouini, M. Cooperative HARQ Assisted NOMA Scheme in Large-scale D2D Networks. *arXiv* **2017**, arXiv:abs/1707.03945.

21. Guo, A.; Zhong, Y.; Zhang, W.; Haenggi, M. The Gauss Poisson Process for Wireless Networks and the Benefits of Cooperation. *IEEE Trans. Commun.* **2016**, *64*, 1916–1929.

22. Yang, Z.; Ding, Z.; Wu, Y.; Fan, P. Novel Relay Selection Strategies for Cooperative NOMA. *IEEE Trans. Veh. Technol.* **2017**, doi:10.1109/TVT.2017.2752264.

23. Ding, Z.; Dai, H.; Poor, H.V. Relay Selection for Cooperative NOMA. *IEEE Wirel. Commun. Lett.* **2016**, *5*, 416–419.

24. Ding, Z.; Dai, L.; Poor, H.V. MIMO-NOMA Design for Small Packet Transmission in the Internet of Things. *IEEE Access* **2016**, *4*, 1393–1405.

25. Sun, Y.; Ng, D.W.K.; Zhu, J.; Schober, R. Multi-Objective Optimization for Robust Power Efficient and Secure Full-Duplex Wireless Communication Systems. *IEEE Trans. Wirel. Commun.* **2016**, *15*, 5511–5526.

26. Zabini, F.; Bazzi, A.; Masini, B.M.; Verdone, R. Optimal Performance Versus Fairness Tradeoff for Resource Allocation in Wireless Systems. *IEEE Trans. Wirel. Commun.* **2017**, *16*, 2587–2600.

27. Ding, Z.; Yang, Z.; Fan, P.; Poor, H.V. On the Performance of Non-Orthogonal Multiple Access in 5G Systems with Randomly Deployed Users. *IEEE Signal Process. Lett.* **2014**, *21*, 1501–1505.

28. Liu, Y.; Ding, Z.; Elkashlan, M.; Yuan, J. Nonorthogonal Multiple Access in Large-Scale Underlay Cognitive Radio Networks. *IEEE Trans. Veh. Technol.* **2016**, *65*, 10152–10157.

29. Men, J.; Ge, J.; Zhang, C. Performance Analysis of Nonorthogonal Multiple Access for Relaying Networks Over Nakagami-*m* Fading Channels. *IEEE Trans. Veh. Technol.* **2017**, *66*, 1200–1208.

30. Hiderband, E. *Introduction to Numerical Analysis*; Dover: New York, NY, USA, 1987.

31. Ding, Z.; Poor, H.V. Cooperative Energy Harvesting Networks With Spatially Random Users. *IEEE Signal Process. Lett.* **2013**, *20*, 1211–1214.

future internet

MDPI

Article

NB-IoT for D2D-Enhanced Content Uploading with Social Trustworthiness in 5G Systems [†]

Leonardo Militano [1] , Antonino Orsino [2,*] , Giuseppe Araniti [1,3] and Antonio Iera [1]

[1] DIIES Deparment, University "Mediterranea" of Reggio Calabria, Reggio Calabria 89100, Italy;
    leonardo.militano@unirc.it (L.M.); araniti@unirc.it (G.A.); antonio.iera@unirc.it (A.I.)
[2] ELT Deparment, Tampere University of Technology, Tampere 33720, Finland
[3] API Deparment, Peoples' Friendship University of Russia (RUDN University), Moscow 101000, Russia
*   Correspondence: antonino.orsino@tut.fi; Tel.: +358-44-299-2908
†   Militano, L.; Orsino, A.; Araniti, G.; Nitti, M.; Atzori, L.; Iera, A. Trusted D2D-based data uploading in in-band
    narrowband-IoT with social awareness. In Proceedings of the IEEE 27th Annual International Symposium on
    Personal, Indoor, and Mobile Radio Communications (PIMRC), Valencia, Spain, 2016; pp. 1–6.

Academic Editor: Boon-Chong Seet
Received: 14 June 2017; Accepted: 6 July 2017; Published: 8 July 2017

Abstract: Future fifth-generation (5G) cellular systems are set to give a strong boost to the large-scale deployment of Internet of things (IoT). In the view of a future converged 5G-IoT infrastructure, cellular IoT solutions such as narrowband IoT (NB-IoT) and device-to-device (D2D) communications are key technologies for supporting IoT scenarios and applications. However, some open issues still need careful investigation. An example is the risk of threats to privacy and security when IoT mobile services rely on D2D communications. To guarantee efficient and secure connections to IoT services involving exchange of sensitive data, reputation-based mechanisms to identify and avoid *malicious* devices are fast gaining ground. In order to tackle the presence of malicious nodes in the network, this paper introduces *reliability* and *reputation* notions to model the level of *trust* among devices engaged in an opportunistic hop-by-hop D2D-based content uploading scheme. To this end, *social awareness* of devices is considered as a means to enhance the identification of trustworthy nodes. A performance evaluation study shows that the negative effects due to malicious nodes can be drastically reduced by adopting the proposed solution. The performance metrics that proved to benefit from the proposed solution are data loss, energy consumption, and content uploading time.

Keywords: trustworthiness; D2D communications; 5G systems; Internet of things; NB-IoT

1. Introduction

The expected drastic increase in Internet of things (IoT) connected devices will definitely produce huge demands for data transmission over wireless systems. At the same time, a plethora of new IoT use cases are emerging across the domains of intelligent transportation systems, smart grid automation, remote health care, smart metering, industrial automation and control, remote manufacturing, and public safety surveillance, among others [1]. Most IoT devices operate through their virtual representations within a digital overlay information system, built over the physical world. Therefore, the majority of current IoT solutions rely on cloud services, leveraging on their virtually unlimited capabilities to effectively exploit the potential of massive tiny sensors and actuators towards the so-called cloud of things. Given the complexity and the challenging requirements of future IoT ecosystems, experts in the field share the opinion that the upcoming fifth generation (5G) cellular systems will represent a strong boost for actual IoT deployment [2]. This vision is sustained by the fervent activities, aimed at designing IoT-oriented 5G wireless systems, conducted by academic, industrial, and standardization bodies [3,4], worldwide. Several types of interactions may coexist within an IoT ecosystem, including machine-to-machine (M2M), machine-to-human, human-to-machine, and machine-to-cloud interactions. All require ubiquitous connectivity. For this purpose, device-to-device (D2D) communications appears as a promising paradigm to support the interconnection of heterogeneous objects [5].

Short-range D2D cooperation among devices may introduce benefits in terms of improved spectrum utilization, higher throughput, and lower energy consumption, which is important for constrained IoT devices. However, there are still several open issues that need to be solved in order to achieve a seamless, effective, and reliable deployment of proximity-based communications for IoT systems [6,7].

For an effective implementation of proximity communications, one of the most important challenges is to understand how the node-originated information shall be processed so as to build a reliable system on the basis of the objects' behavior, namely the need for *trustworthiness* [8]. Indeed, in realistic scenarios, where human interactions and human behavior are also to be considered, the presence of *malicious* nodes in the network is a constant threat for successful cooperation. Accordingly, without effective trust management foundations, attacks and malfunctions are likely to outweigh any possible cooperation benefits [8].

For the reference scenario in this paper, we consider that groups of devices in close proximity are willing to upload contents to the Cloud or to a central server and end users may not be aware of whom they are going to forward the data to. Typical sample scenarios are small-scale crowded environments (e.g., stadiums, university campuses, music theaters, or fairs) where devices can exploit opportunistic data forwarding over other devices in proximity. In these contexts, malicious nodes may decide to drop the data packets they are expected to forward or even modify the data packets before forwarding the corrupted content. To cope with these threats, *reliability* and *reputation* notions will be considered to model the level of *trust* among the involved entities.

By taking inspiration from recent social Internet of things (SIoT) models, in this paper we consider the sociality level of the devices to model the *reliability* of the communication. The historical *reputation* of the cooperative users will be considered to offer rational users the possibility to filter out untrusted users and avoid unsuccessful opportunistic hop-by-hop D2D interactions. An initial investigation in this direction was made in our previous paper [9] in long-term evolution-advanced (LTE-A) scenarios where multihop cooperative uploading is implemented over cellular D2D resources [10]. In this paper we take forward our research, investigating among other issues the use of the recent narrowband IoT (NB-IoT) standard [11], which is currently considered the reference cellular technology for IoT communications for the next 5G systems. The *trust* constraints for successful D2D-based content uploading are modeled by including sociality among devices, as a measure of *reliability*, and historical *reputation*. The objective is to define multihop D2D topologies that meet the constraints of reciprocal user equipment (UE) proximity for the direct links activation and, at the same time, of an adequate *trust* level among the cooperating devices. Through simulation-based performance evaluations, we show that it is possible to significantly reduce the impact of malicious behaviors on the performance of involved devices, with gains in terms of data loss, energy consumption, and data uploading time.

The remainder of the paper is organized as follows: Section 2 reviews the related work; Section 3 introduces the research background and motivation; the algorithmic solution for the definition of trusted D2D cooperative topologies is given in Section 4; a detailed description of the proposed trust model and the sociality concepts is given in Section 5; and numerical results and conclusions are provided in Section 6 and Section 7, respectively.

2. Related Work

Security is one of the key issues for an effective and widespread adoption of D2D communications [12] in IoT scenarios [13]. This is particularly relevant in a cooperative context such as the one studied in this paper, where the multihop D2D data forwarding paradigm is based on the assumption that the involved devices behave in a trusted and secure way [14]. Unfortunately, this is not always the case as *malicious* nodes may be active in the network by either dropping or manipulating the data to be forwarded.

Generally, trust is defined as the quantified belief of a truster with respect to the competence, honesty, security and dependability of a trustee within a specified context [15]. When two users want to cooperate, one of them (the truster) assumes the role of a service requester and the other (the trustee) acts as the service provider. Specifically, in our cooperative D2D multihop scenario, the node acting as relay/gateway towards another node will be the trustee and the source node of the relayed data is the truster. The cooperative topology formation exploits a game theoretic coalition formation model as proposed in [10]. Game theoretic approaches have found several applications also in the field of D2D communications given the potential to model the user behavior (see e.g., [16,17]). The trustworthiness of the truster with respect to the trustee can be determined by considering reliability and/or reputation. The former is a direct measure derived by subjective observations of the truster during its interactions with the trustee; the latter is an indirect measure based on the opinions that other actors in the community have about the trustee.

In the literature, several trust models have been proposed to represent both reliability and reputation [15]. A way to reach trustworthiness in communication is to exploit sociality among devices [9,18]. The mechanism we propose enhances classic trust models through the exploitation of *social relationships* among the involved devices (to improve device *reliability*) and of recommendation exchange (to the purpose of reputation definition). Socially-aware D2D communications have attracted high interest in recent research activity, such as for instance in [19–22]. With respect to the works in the literature, we consider the potential of the SIoT model defined in [23], to embrace the social networking concepts and build trustworthy relationships among devices [24,25]. In particular, mobility patterns and relevant context can be considered to configure the appropriate forms of socialization among the UE. Specifically, the so-called *co-location object relationships* (C-LOR) and *co-work object relationships* (C-WOR) are established between devices in a similar manner as among humans, when they share personal (e.g., cohabitation) or public (e.g., work) experiences. Another type of relationship may be defined for the objects owned by a single user, which is named *ownership object relationship* (OOR). The parental object relationship (POR) is defined among similar devices built in the same period by the same manufacturer, where the production batch is considered a family. Finally, the social object relationship (SOR) is established when objects come into contact, sporadically or continuously, for reasons related to relations among their owners.

3. NB-IoT and D2D Communications in the 5G Era

The upcoming fifth generation (5G) wireless systems are being considered as the best candidate to allow effective interworking of IoT devices, thanks to the benefits these offer in terms of enhanced coverage, high data rate, low latency, low cost per bit, and high spectrum efficiency. There is a general consensus among academia and industries that 5G will have a huge impact in three main areas of communication: (1) enhanced mobile broadband (eMBB); (2) massive-machine type communication (M-MTC); and (3) critical-MTC (c-MTC). In particular, these three areas have different requirements and applications. Nevertheless, those are not standalone use cases, but may overlap in some cases. In our work, we take into consideration a use case that is somehow in the middle between c-MTC and M-MTC. We may think of process automation within a factory or other similar scenarios. In these cases, available and reliable connections for monitoring and diagnosis of a high number of industrial elements (i.e., M-MTC) are the most important. Nevertheless, even if the measured values from the sensors change relatively slowly, it is still important to have reasonable latency (e.g., from 20 to 50 ms) in order to react in a timely manner to an issue that can occur on the way (e.g., c-MTC).

With reference to typical machine-type communications (MTC), the Third Generation Partnership Project (3GPP) has introduced novel features [26] that better support the intrinsic battery-constrained capabilities of IoT devices and the typical small data packets over licensed bands (e.g., LTE). In September 2015, 3GPP standardized *narrowband IoT (NB-IOT)*, a new narrowband radio technology to address the requirements of the Internet of things (IoT). This new technology provides improved indoor coverage, support of massive number of low throughput devices, has low delay sensitivity, ultra-low device cost, low device power consumption and an optimized network architecture.

At the time we are writing this paper, a first release of NB-IoT had been completed by 3GPP. However, the standardization process is still ongoing and further enhancements and new features are expected in 3GPP Release 14 (updated according to the last 3GPP meetings) and Release 15. Further, NB-IoT is expected to be released in a form of a software update for the network operators and is fully backward compatible with existing 3GPP devices and infrastructure. In particular, given an available bandwidth of around 200 kHz for both downlink and uplink, the air interface of NB-IoT is optimized to ensure harmonious coexistence with LTE. In particular, the technology can be deployed "in-band" using the resource blocks within a normal LTE carrier (an LTE operator can deploy NB-IoT inside an LTE carrier by allocating one of the physical resource blocks (PRB) of 180 kHz to NB-IoT), or in the unused resource blocks within a LTE carrier guard-band (for instance, for an LTE bandwidth of 10 MHz (i.e., 56 resource blocks—RBs), 6 RBs are reserved for guard subcarriers and can be used for NB-IoT), or in "standalone" manner for deployments in dedicated spectrum [27]. Thanks to this latter feature whereby NB-IoT may be deployed as a stand-alone carrier using any available spectrum exceeding 180 kHz, a Global System for Mobile Communications (GSM) operator can also replace a GSM carrier (200 kHz) with NB-IoT. As reported in the white paper from Nokia [11], the maximum data rates (i.e., by considering the overall bandwidth) in terms of instantaneous peak rates provided by the NB-IoT technology are: *170 kbps (Downlink – DL)* and *250 kbps (Uplink – UL)*. To enable the allocation of small portions of bandwidth, NB-IoT uses tones or subcarriers instead of resource blocks. The subcarrier bandwidth for NB-IoT is 15 kHz (or 3.75 kHz in some cases), compared with a resource block, which has an effective bandwidth of 180 kHz. Furthermore, the data rates available for the single tone in downlink and uplink are 680 bits and 1000 bits, respectively. These values will satisfy most of the communication requirements for IoT-based services where very small data packets are usually transferred.

Another form of technology which has gained high momentum in the evolution towards 5G systems is D2D communication where devices communicate directly over cellular resources or Wi-Fi/Bluetooth technologies without routing the data over a base station (BS) or an access point (AP). Recent studies showed how D2D communications may find important applications in IoT/5G integration [6,7]. Indeed, D2D communications not only allow for extending the coverage and overcoming the limitations of conventional cellular systems, but they represent a fertile ground for use cases and services (e.g., social interactions and gaming, local information exchange, etc.). For instance two users can find each other whenever in proximity and share data or play interactive games. Moreover, social applications, public safety and emergency handling may benefit from D2D communications as devices can provide local connectivity in case of damage to the network infrastructure. Other fields of applications may be vehicle-to-vehicle (V2V) communication in intelligent traffic systems (ITS) where D2D communications can be exploited for traffic control/safety applications among others.

Several works in the recent literature have investigated the benefits D2D communications can introduce, making it a very appealing solution for the exacting requirements of IoT emerging 5G network scenarios [28,29]. The most important of these benefits are [30]: (1) higher data rate in the communications; (2) reliability in the communications including in the case of network failure; (3) energy savings due to lower transmission power levels for devices in proximity; (4) reduced number of cellular connections (known as traffic offloading); and (5) possibility for instantaneous communications between devices.

In this paper, the potential benefits of NB-IoT and D2D communications are jointly exploited for cooperative content uploading from a set of devices to the cellular base station, through short-range multihop relaying. In particular, NB-IoT is exploited for radio links between users and the eNodeB, whereas proximity-based transmissions (i.e., D2D) are established among devices in mutual proximity.

However, a necessary condition for such a "cooperative" relaying solution to bring benefits compared to the "non-cooperative" case, is that the link quality of the multihop D2D channels is higher than the one of the separate links to the Internet. This condition is more likely to occur in non-isotropic propagation environments with obstacles where non-line-of-sight (NLOS) conditions may cause partial and temporary out-of-coverage conditions, as is the case of the Internet of vehicles, an instance of the IoT where objects are represented by cars [31]. The content we have in mind for the devices in the scenario is small data coming for instance from sensing activities, security or monitoring applications with limited amount of data to transmit, typical of IoT applications. NB-IoT is, in fact, not thought of for bandwidth-hungry applications, e.g., videos, or big file transmissions. Currently IoT devices are equipped with a wide range of radio technologies. For instance, Pycom (https://www.pycom.io/) provides some shields for IoT applications that include both long- and short-range connectivity such as LoRa, LTE, NB-IoT, Bluetooth, and LTE Cat-M1. The idea we want to investigate is to offload the part of the traffic that cannot be handled entirely by NB-IoT via short-range links over the D2D technology.

4. Cooperative Multihop D2D-Based Data Uploading

We consider a single LTE-A cell with multiple devices interested in uploading their content to the Internet by adopting an *in-band NB-IOT* solution. Data uploading according to the traditional *cellular-mode* is performed through the activation of separate links from each device to the eNodeB. The alternative solution proposed in this paper is the cooperative content uploading controlled by the eNodeB (i.e., network-assisted D2D), where the UE organizes itself to form a "logical multihop D2D topology" and cooperates in uploading the content generated by *all* the involved devices. The cooperative topology formation is implemented according to a game theoretic coalition formation model as proposed in [10].

In the formed cooperative topology, the user equipment (UE) located farther from the base station relays its content to nearby UE and only the UE playing the head-end role in the topology, the so-called *gateway*, is in charge of uploading all the contents received from the rest of the UE to the eNodeB. The UE with the best link quality in the coalition is chosen as the gateway and may receive (if needed) all the radio resources that would have been separately allocated by the eNodeB to the UE in the coalition. Of course, since NB-IoT technology is used, in this case the radio resources are "tones" rather than the classic definition of resource blocks (RBs). For example, a channel bandwidth of 20 Mhz corresponds to 100 RB of LTE-A. The RB corresponds to the smallest time frequency resource that can be allocated to a user (12 sub-carriers) in LTE. The intermediate UE in the topology also acts as *relays* for the contents received from the upstream UE. In doing this, they benefit from the higher quality of the short D2D links with respect to the direct cellular link. In the most general configuration, each relay has one or more links active to receive data from the preceding sources in the topology, and one single link active to relay data (its own generated traffic and the traffic from the incoming D2D links) to the subsequent UE in the topology. Each UE operates in half-duplex mode; thus, it either receives or transmits in a given transmission time interval. We consider a reasonable assumption for *rational* self-interested devices, that each UE uploads its own generated content first and then the content received by the preceding UE in the topology. In particular, the transmission starts only after the generic UE has received the whole content (in other words, UE uses the decode-and-forward relaying protocol).

We remark that all the transmission between one single device towards the eNodeB exploits the NB-IoT tones, whereas D2D links are activated over the legacy LTE spectrum (thus using RBs instead of tones). The motivation of this choice is driven by the fact that NB-IoT does not yet support proximity-based transmission even if this feature is actually been discussing during the 3GPP Release 15 standardization.

In realistic scenarios, end-users may not be aware of whom they are going to be connected to and *malicious* devices may decide to either modify the received content before forwarding a corrupted packet or drop data packets they are expected to forward without informing the interested users.

In the remainder of the paper we will refer to these two different types of malicious nodes as *type A* and *type B*, respectively. Whenever a malicious node in the coalition either modifies or drops the data packets, we assume that the source node will be informed by the eNodeB and will perform standard data uploading. This will introduce both a delay in the data delivery and an increase in the energy consumption for the involved nodes. However, there is a difference, since the effect of a *type B* malicious node is identified earlier than the effect of a *type A* (note that type A malicious nodes are present only when unencrypted payloads are delivered) node. In fact, if a timeout is enough to identify a packet dropping, a corrupted packet will first have to reach the eNodeB passing through all the intermediate nodes before the system is aware of data corruption (we assume the eNodeB will be able to identify data corruption).

We imagine a possible way for the eNodeB (eNB) to detect a packet to be dropped (when a *type B* malicious node is present). A straightforward solution is to rely on the acknowledgment message (ACK) that has to be sent to the UE once the packet has been received by the network. In this case, once the UE reaches the maximum number of re-transmitted Protocal Data Units (PDUs), a radio link failure (RLF) is triggered and sent over the signal radio bearers (SRBs). Since in our work we assume that malicious nodes of type B drop the packets, we think that the best way to proceed is for the eNB to send the ACKs directly to each UE of the chain. Then, if a RLF is experienced (due to no ACK being received, i.e., the gateway does not forward anything), this is triggered separately by each UE towards the eNB. The identification of a corrupted packet is a bit more complex. This may be done through the checksum of the packets (i.e., at the higher layer of the protocol stack) or, alternatively, through the integrity protection that at the moment is done for the SRBs. We are aware that integrity protection is not present for the data radio bearer (DRBs), but with the ongoing standardization of 5G New Radio (NR) and LTE Evolution (LTE-Evo) we may expect this kind of enhancements or new features. Other solutions may be applied for detecting a not correct node behavior, but this is out of the scope of this work.

To cope with the threats coming from the malicious nodes, when defining the cooperative topologies, countermeasures must be considered to offer rational users the possibility to filter out untrusted users, block the unsecure links and avoid unsuccessful opportunistic hop-by-hop D2D interactions, as sketched in Figure 1. The solution we propose for effective and trusted D2D-based data uploading can be summarized as follows:

Channel quality indicator (CQI) collection: the eNodeB collects the CQI values from each unit of UE, relevant to the direct links with all its neighbors and to the uplink toward the eNodeB.

Virtual resource allocation: the eNodeB considers the situation where the single UE devices are transmitting in unicast over their uplink and computes the radio resources according to the scheduling policy. The so-computed radio resources are considered as "virtual" since they are not yet allocated to the UE because the UE may actually form a cooperative multihop topology (i.e., a coalition). Whenever a coalition is formed, the pool of "virtual" resources of all the UE in the coalition will be assigned to the respective gateway.

Cooperative coalition formation: in this step a set of stable coalitions are determined where for each coalition the roles for the nodes in the cooperative D2D-based data uploading are defined, as well as the routing path for the data from each node. To produce stable coalitions, the eNodeB will rely on a game theoretic model such as the one defined in [10]. A classic merge and split algorithm is implemented where the device preference to join or leave a coalition is based on the estimated data uploading time. In the coalition formation algorithm two main constraints are considered: (1) two consecutive nodes in the data routing path built on a cooperative coalition must be in coverage for a D2D link (otherwise the data routing would fail); and (2) the devices in the cooperative coalitions should guarantee a minimum value of trust which we define as *feasibility threshold FT*. Indeed, we consider a coalition as *not feasible* if at least one link $i \rightarrow j$ in the topology does not meet the constraint:

$$pt_{i,j} \cdot d_{i,j} \geq FT \qquad (1)$$

where $pt_{i,j} \rightarrow [0,1]$ is the *player trust* that player i (a device in a coalition) associates to player j (see Section 5). The second term $d_{i,j}$ is a binary function taking value 0 if users i and j are not in proximity, and value 1 otherwise. A link not meeting the mentioned constraint is represented in Figure 1 as a blocked link.

Data transmission configuration: For each coalition that is formed, the eNodeB assigns the respective pool of virtual radio resources to the gateway, and transmits all the required information to the UE so that the transmissions can start. The devices in different coalitions are always allocated to orthogonal frequency resources by the scheduler (we consider a *maximum throughput* scheduler) so that mutual interference is avoided. The configuration of the D2D communications assumes that UE simultaneously transmitting in the same coalition adopts different RBs to avoid any mutual interference (this leads to a worst case analysis and better results can be obtained with enhanced interference management).

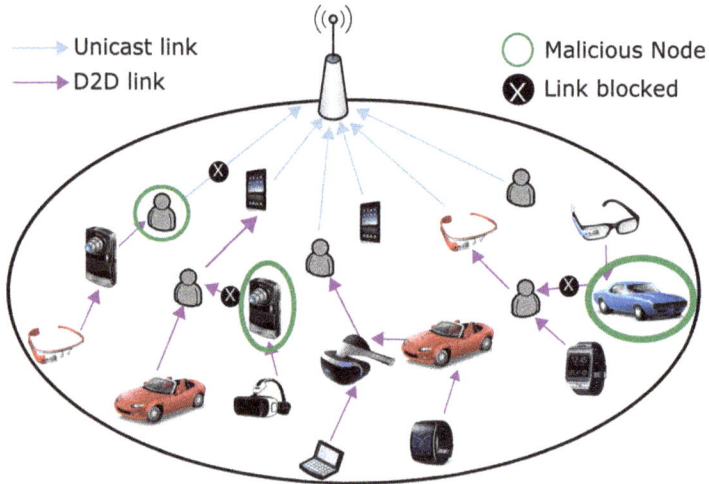

Figure 1. Cooperative multihop content uploading based on trustworthy device to device (D2D) links.

5. The Social-Aware Trust Model

In our scenario the eNodeB acts as a trusted third party that implements the coalition formation model based on social-aware trustworthiness. To this aim, we evaluate the potential of the SIoT model to embrace the social networking concepts and build trustworthy relationships among the devices [25]. The eNodeB will store information about the reliability, reputation and trust of the users in the network. We define a *player trust matrix (PTM)* as the data structure stored in the eNodeB containing information for every pair of devices. This information will be used whenever a new coalition formation is triggered. Every element (i-th row and j-th column) of the PTM refers to a D2D link connecting the corresponding $i \rightarrow j$ nodes in the coalition being considered at time t, where node j is the relay/gateway for the data he receives from node i (its own and the preceding nodes in the topology); we consider i, j as the truster and the trustee respectively. The eNodeB will also act as a controller of the data uploading success as it will send an acknowledgment to the respective source nodes after each cooperative data transmission. Whenever data loss or data corruption is detected by the eNodeB, malicious behavior will be detected and the information about the reliability of the interested devices will be respectively updated. We assume that control messages (sent over control channels) are very small compared to the main content to be sent and therefore, the relative transmission time and energy consumption are assumed to be negligible. The parameters used to define the level of trust are the following:

Social player reliability ($spr_{i,j}$): this parameter has a value in $[0,1]$ and measures the reliability that node *i* assigns to player *j* based on the social relationship between the two devices;

Player reliability ($pr_{i,j}^t$): this parameter has a value in $[0,1]$ and is representative of the reliability at time instant *t* that player *i* assigns to player *j*. To determine this value, each player will consider both the *social player reliability* and the outcome of past interactions (only those cases are considered where player *j* was expected to act as relay/gateway for player *i*).

Recommendation reliability ($rr_{i,j}$): this parameter has a value in $[0,1]$ and is a measure of the reliability assigned by node *i* to the recommendations it receives from another device *j* about third devices in the network. In our model we consider this parameter to be influenced by the social relationship between the interested UE.

Player reputation ($pp_{i,j}^t$): this parameter has a value in $[0,1]$ and measures the reputation of player *j* for player *i* according to the information he received through the recommendation values from third players in the network at a given time instant *t*.

Player trust ($pt_{i,j}^t$): this parameter has a value in $[0,1]$ and measures the level of trust for player *j* at time *t* as evaluated by player *i*. This is the most important parameter as it will determine whether player *i* is willing to consider player *j* as relay/gateway node in a D2D-based cooperative coalition. For the computation of its value player *i* will use a weighted combination of the reliability $pr_{i,j}^t$ and the reputation $pp_{i,j}^t$ parameters.

As commented above, the *player reliability* parameter is a function of the time instant *t*. In particular, its value is updated at every time instant based on the experienced behavior of the devices in the cooperative data uploading. To make this work, for each cooperative interaction the eNodeB sends an acknowledgment to the source nodes with information about the data being successfully received. Of note, this does not allow to determine which node in the cooperative topology has actually dropped or corrupted the data. Therefore, we assume that the eNodeB will associate the outcome value δ_d to the node *j* that was entrusted by node *i* as relay/gateway forming a D2D link $i \rightarrow j$. At time instant $t = 0$ the only information the interested devices can exploit for judging the *player reliability* is *social player reliability* ($spr_{i,j}$) which is set according to predefined values (see Table 1). If two communicating entities are tied by two or more types of relationships, the strongest tie with the highest factor has to be considered [25]. At subsequent time instants $t > 0$, the results of cooperative interactions can be used to determine the player reliability $pr_{i,j}^t$ with *j* acting as relay/gateway for data sent by *i*. In particular, we define with $\Delta_{i,j}^t = \{\delta_1, \ldots, \delta_d \ldots \delta_D\}$ the set of past interactions registered until time *t*, where the generic $\delta_d \in [0,1] \in \mathbb{R}$ is equal to the total percentage of data that has been successfully forwarded by node *j* and reached the eNodeB. Summarizing, we define the player reliability $pr_{i,j}^t$ as follows:

$$
pr_{i,j}^t = \begin{cases} spr_{i,j} & t = 0 \\ \alpha \cdot spr_{i,j} + (1-\alpha) \cdot \dfrac{\sum_{d \in \Delta_{i,j}^t} \delta_d}{|\Delta_{i,j}^t|} & t > 0 \end{cases} \tag{2}
$$

where $\alpha \in [0,1]$ is a weighting factor to give more or less importance to the initial sociality relationship between the nodes.

The other parameter that is being updated after each cooperative interaction is the player *reputation* which is based on the opinions of the community in the network. If, for instance, a player *i* asks the opinion about player *j* to the community, it will receive an opinion from a set of players in the network. Let us say this set of players is $\mathcal{K} \subseteq \mathcal{N} \setminus \{i\}$, where $\mathcal{N}$ is the total set of devices in the network. The opinion player *k* will provide is its own measure of trust about player *j* at time instant *t*, namely $pt_{k,j}^t$.

Table 1. Player and recommendation reliability values associated to the social relationship between devices.

Relationship	Description	Social player Reliability ($spr_{i,j}$)	Recommendation Reliability ($rr_{i,j}$)
Ownership object relationship (OOR)	Objects owned by the same person	1	0.9
Co-location object relationship (C-LOR)	Objects sharing personal experiences	0.8	0.6
Co-work object relationship (C-WOR)	Objects sharing public experiences	0.7	0.5
Social object relationship (SOR)	Objects in contact for owner's relations	0.6	0.5
Parental object relationship (POR)	Objects with production relations	0.5	0.4
No relationship		0.1	0.1

To best weigh the opinions received from the other players, a confidence factor called *recommendation reliability* ($rr_{i,k}$) is used. In our proposed model this is set according to the social relationship between the involved devices as reported in Table 1. Note that we assumed the *recommendation reliability* to have a lower value with respect to *social player reliability* in general. The motivation for this is that the recommendation received by a socially related device may be influenced by the outcome of past cooperative iterations which affected the ability to provide an objective recommendation. Given the collected information, the *player reputation* at time *t* is computed as follows:

$$pp_{i,j}^t = \frac{\sum_{k \in \mathcal{K}} rr_{i,k} \cdot pt_{k,j}^t}{\sum_{k \in \mathcal{K}} rr_{i,k}} \tag{3}$$

Player *i* can then determine the player trust value $pt_{i,j}^t$ it associates to player *j* at time instant *t*, as a combination of the player reliability ($pr_{i,j}^t$) and the player reputation ($pp_{i,j}^t$) weighted by a real coefficient β ranging in $[0,1] \in \mathbb{R}$:

$$pt_{i,j}^t = \begin{cases} 0.5 & t = 0 \\ \beta \cdot pr_{i,j}^t + (1-\beta) \cdot pp_{i,j}^t & t > 0 \end{cases} \tag{4}$$

The choice to set the initial trust value to 0.5 is caused by *whitewashing strategies* where a malicious adviser can whitewash its low trustworthiness starting a new account with the initial trustworthiness value.

6. Performance Evaluation

In this section we provide the output of an extensive simulation campaign finalized to demonstrate the robustness of the proposed solution to the presence of malicious nodes. The presented results are obtained using a built-in simulator in Matlab already used in previous works [9,10]. The proposed solution, hereafter named *trust-based*, is compared to an alternative *basic* approach that does not take into account any trustworthiness for the involved users and is unable to detect the malicious behavior. As discussed earlier (see Section 4), we consider two different types of malicious nodes, i.e., (1) *type A*, where users forward corrupted packets (for instance) to perform an attack to security, and (2) *type B*, where users drop the packets to exploit the benefits given by multi-hop D2D connections without forwarding any content further in the chain.

The reference scenario is composed by a single LTE-A cell with a 500-m radius and 10-MHz bandwidth (i.e., 50 RBs available) where 20 UE devices are uniformly distributed. As for the NB-IoT, we use the "in band" where 6 RBs (for a total number of 288 tones) are allocated for the transmissions among the selected gateways and eNodeB. The main simulation parameters are listed in Table 2. The content size for all the nodes is set to 50 MB and radio resources used on a D2D transmission are limited to the so-called "virtual resources" allocated by the eNodeB to the involved pairs of UE (see Section 4 for more details). The performance parameters we focus on for the system-level performance are: (1) *data loss*; (2) *average data uploading time gain*; and (3) *average energy consumption gain*.

In particular, the latter two parameters represent the gain achieved by the cooperative upload a pure cellular upload solution where each user uploads directly the content to the network infrastructure by using standard LTE unicast transmissions.

Table 2. Main simulation parameters. NB-IoT: narrowband Internet of things; CQI: channel quality indicator; MCS: modulation and coding scheme; TTI: Transmission time interval; TDD: Time division duplex.

Parameter	Value
Cell radius	500 m
Maximum D2D link coverage	100 m
TTI	1 ms
TDD configuration (D2D)	0
Carrier frequency	2.1 GHz
Tx Cellular power (NB-IoT)	23 dBm
Tx D2D power	-19 dBm
CQI-MCS mapping for D2D links	"refer to [32]"
Noise power	-174 dBm/Hz
Cellular link model	Rayleigh fading channel
D2D link model	Rician fading channel
NB-IoT tones	288 (i.e., 6 RBs)
Content size	50 MB
Weighting factors $\alpha = \beta$	0.5
Simulation time	100 s
# of Runs	500

The first analysis we discuss is the impact that the two classes of malicious nodes have on the uploading time and the energy consumption. In particular, we consider three different distributions of malicious nodes in the system, namely: (1) prevalence of type A (i.e., 75–25%) malicious node; (2) equal number of type A and type B malicious nodes (i.e., 50–50%); and (3) prevalence of type B malicious nodes (i.e., 25–75%). As we can observe from Figure 2, the proposed trust-based solution always performs better compared to the basic strategy (we consider here the sample case with FT = 0.5). In particular, when there is a prevalence of type A malicious nodes in the system we obtain lower benefits in terms of uploading time and energy consumption. The motivation behind this is that the energy consumed for a UE when receiving corrupted packets is added to the energy required to upload the content with a unicast link to the eNodeB. In the presence of type B malicious nodes (i.e., dropping packets), instead, the only energy consumption for the UE is due to the unicast uplink transmission from the UE to the eNodeB. Same motivations yield for the differences observed in the uploading time gain, which, as shown in Figure 2a, results to be higher when we have a prevalence of type B malicious nodes.

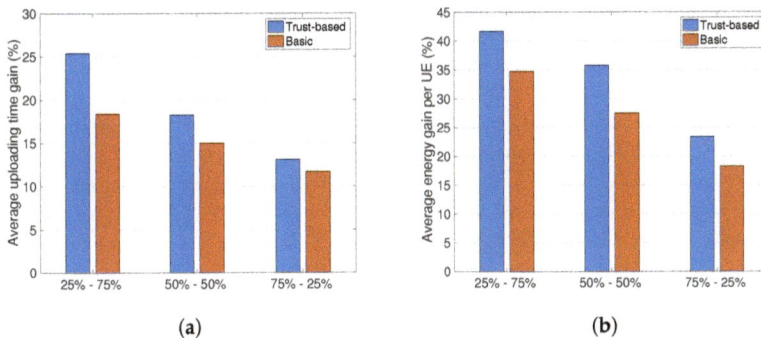

Figure 2. Impact of type A and type B malicious nodes (feasility threshold, FT = 0.5). (a) Uploading time gain; (b) Average energy gain.

The next analysis shows the results when the percentage of malicious nodes in the system varies in the range [15–90%]. Here it is assumed that there is an equal number of malicious nodes of type A and type B and FT = 0.5. As we can observe from the plots in Figure 3a,b, the proposed trust-based solution obtains better performance. In particular, the average uploading time gain and the energy consumption gain are higher with the proposed solution. In fact, with our approach users forward data to trusted devices in proximity by avoiding transmissions with malicious nodes. In details, the achieved gain compared to the basic solution reaches the value of +5% and +7% (on average) for uploading time and energy consumption, respectively. This behaviour is also confirmed by curves in Figure 3c showing the amount of data loss due to malicious nodes. Here, the trust-based solution has a percentage of data loss that is 19% (on average) less than the data loss with the basic approach.

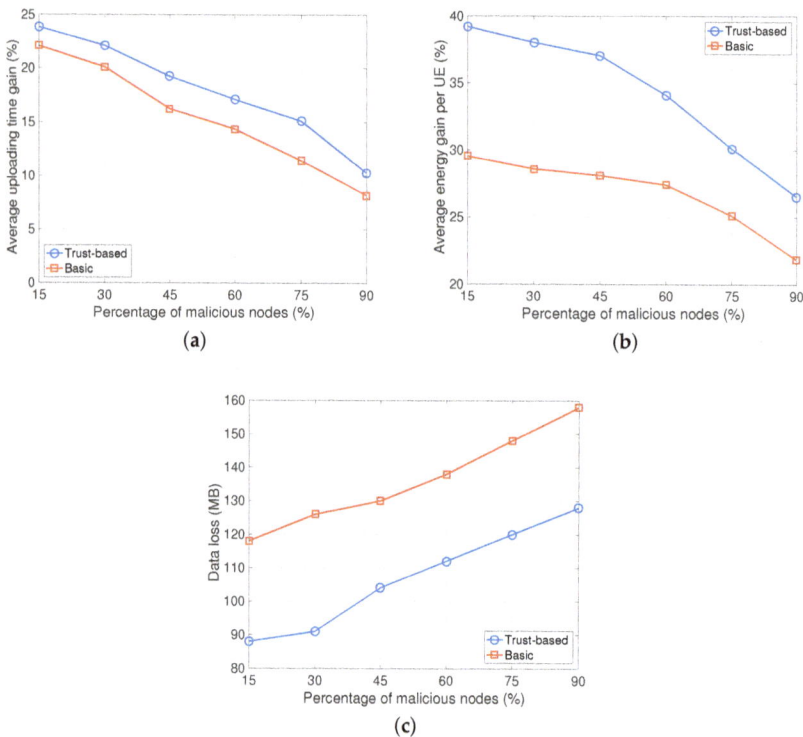

(a) (b)

(c)

Figure 3. Impact of malicious nodes percentage (half of type A and half of type B malicious nodes are considered, FT = 0.5). (a) Uploading time gain; (b) Average energy gain; (c) Data loss.

The last analysis has the objective to show the effects of the *feasibility threshold* on the system-level performance. In Figure 4 results are presented when varying the FT value from 0.2 to 1.0, under the condition of 50% malicious nodes in the system. Interestingly, the gain achieved in terms of uploading time increases linearly with the value of FT until reaching a value of 31% (see Figure 4a). However, this result is obtained at the cost of a higher energy consumption for the nodes. As shown in Figure 4b, the energy consumption gain decreases with the FT and the proposed solution performs even worse than the basic one for FT values beyond 0.5. The reason is that the devices select only nodes with high trustworthiness to forward data. For this reason, the selection of the links to forward the data is strongly constrained and it may be that transmissions occur over low capacity links which require more energy. However, users are able to upload their data without requiring additional transmissions toward the eNodeB. In the extreme case, when the feasibility threshold is set to 1 the amount of data loss is about 19 MB compared to 120 MB for the basic solution.

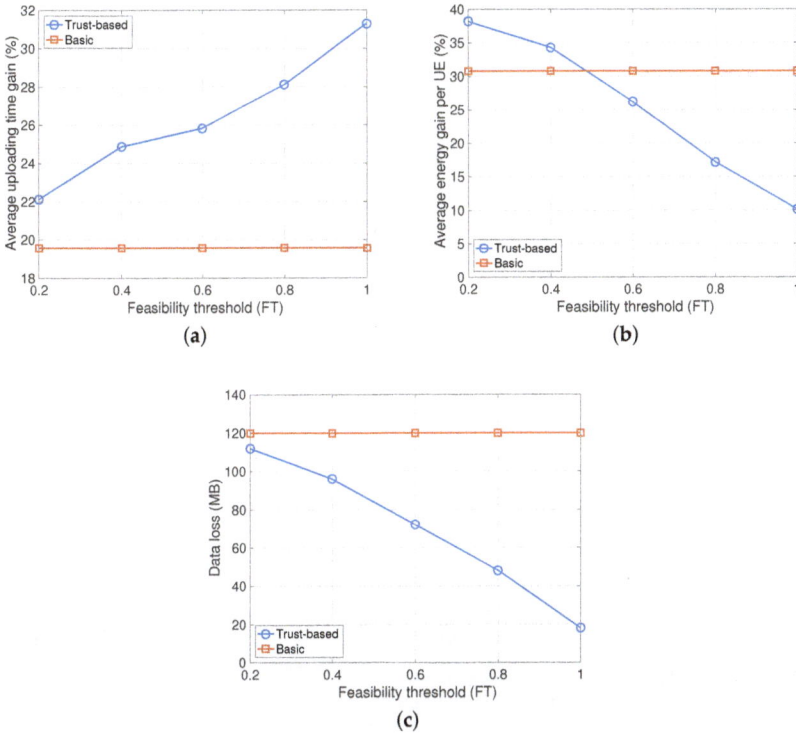

Figure 4. Impact of feasibility threshold (50% of malicious nodes, half of type A and half of type B). (a) Uploading time gain; (b) Average energy gain; (c) Data loss. UE: user equipment.

In conclusion, the proposed trust-based approach outperforms the basic solutions. Moreover, the highest benefits are obtained by tuning the feasibility threshold to the estimated number of malicious nodes in the system and to the desired performance parameter. In fact, even if a high feasibility threshold increases the chances of correctly forwarding the data towards to eNodeB with lower uploading times and data losses, this can result in a higher energy consumption.

7. Conclusions

In this paper we proposed a trust-based solutions for effective D2D-enhanced cooperative content uploading in narrowband-IoT cellular environments. To limit the impact of the malicious nodes either dropping or corrupting the data packets in a cooperative multihop coalition, social awareness has been modeled to evaluate the reliability for the nodes and to suitably weigh the recommendations exchange for the reputation definition. A simulative analysis validated the proposed solution in a wide range of settings for small-scale IoT scenarios. The results showed how the social-based trusted solution guarantees higher gains in the content uploading time, in the energy consumption, and has the ability to increase the amount of successful cooperative interactions by filtering out the malicious nodes.

Acknowledgments: The publication was financially supported by the Ministry of Education and Science of the Russian Federation (the Agreement number 02.a03.21.0008).

Author Contributions: In this paper, the first three authors Antonino Orsino, Giuseppe Araniti and Leonardo Militano conceived the idea, organized the work and designed the analytical model and the proposed algorithms; Antonino Orsino and Leonardo Militano conceived and designed the experiments, analyzed the results and wrote the paper; Antonino Orsino performed the experiments; Antonio Iera supervised the work; all the authors reviewed the writing of the paper, its structure and its intellectual content.

Conflicts of Interest: The authors declare no conflict of interest.

References

1. Fantacci, R.; Pecorella, T.; Viti, R.; Carlini, C. A network architecture solution for efficient IOT WSN backhauling: Challenges and opportunities. *IEEE Wirel. Commun.* **2014**, *21*, 113–119.
2. Soldani, D.; Manzalini, A. Horizon 2020 and Beyond: On the 5G Operating System for a True Digital Society. *IEEE Veh. Technol. Mag.* **2015**, *10*, 32–42.
3. Sachs, J.; Beijar, N.; Elmdahl, P.; Melen, J.; Militano, F.; Salmela, P. Capillary networks: A smart way to get things connected. *Ericsson Rev.* **2014**, *8*, 1–8.
4. Andreev, S.; Galinina, O.; Pyattaev, A.; Gerasimenko, M.; Tirronen, T.; Torsner, J.; Sachs, J.; Dohler, M.; Koucheryavy, Y. Understanding the IoT connectivity landscape: A contemporary M2M radio technology roadmap. *IEEE Commun. Mag.* **2015**, *53*, 32–40.
5. Boccardi, F.; Heath, R.W.; Lozano, A.; Marzetta, T.L.; Popovski, P. Five disruptive technology directions for 5G. *IEEE Commun. Mag.* **2014**, *52*, 74–80.
6. Bello, O.; Zeadally, S. Intelligent Device-to-Device Communication in the Internet of Things. *IEEE Syst. J.* **2014**, *10*, 1–11.
7. Militano, L.; Araniti, G.; Condoluci, M.; Farris, I.; Iera, A. Device-to-Device Communications for 5G Internet of Things. *EAI Endorsed Trans. Internet Things* **2015**, *15*. doi:10.4108/eai.26-10-2015.150598.
8. Roman, R.; Najera, P.; Lopez, J. Securing the internet of things. *Computer* **2011**, *44*, 51–58.
9. Militano, L.; Orsino, A.; Araniti, G.; Nitti, M.; Atzori, L.; Iera, A. Trusted D2D-based data uploading in in-band narrowband-IoT with social awareness. In Proceedings of the IEEE 27th Annual International Symposium on Personal, Indoor, and Mobile Radio Communications (PIMRC), Valencia, Spain, 4–8 September 2016; pp. 1–6.
10. Militano, L.; Orsino, A.; Araniti, G.; Molinaro, A.; Iera, A. A Constrained Coalition Formation Game for Multihop D2D Content Uploading. *IEEE Trans. Wirel. Commun.* **2016**, *15*, 2012–2024.
11. Nokia. LTE Evolution for IoT Connectivity White Paper. In Nokia White Paper. Available online: https://resources.ext.nokia.com/asset/200178 (accessed on 6 December 2016).
12. Gandotra, P.; Jha, R.K.; Jain, S. A survey on device-to-device (D2D) communication: Architecture and security issues. *J. Netw. Comput. Appl.* **2016**, *78*, 9–29.
13. Sicari, S.; Rizzardi, A.; Grieco, L.; Coen-Porisini, A. Security, privacy and trust in Internet of Things: The road ahead. *Comput. Netw.* **2015**, *76*, 146–164.
14. Ometov, A.; Orsino, A.; Militano, L.; Moltchanov, D.; Araniti, G.; Olshannikova, E.; Fodor, G.; Andreev, S.; Olsson, T.; Iera, A.; et al. Toward trusted, social-aware D2D connectivity: Bridging across the technology and sociality realms. *IEEE Wirel. Commun.* **2016**, *23*, 103–111.
15. Grandison, T.; Sloman, M. Trust management tools for internet applications. In *Trust Management*; Springer: Berlin, Germany, 2003; pp. 91–107.
16. Antonopoulos, A.; Kartsakli, E.; Verikoukis, C. Game theoretic D2D content dissemination in 4G cellular networks. *IEEE Commun. Mag.* **2014**, *52*, 125–132.
17. Antonopoulos, A.; Verikoukis, C. Multi-player game theoretic MAC strategies for energy efficient data dissemination. *IEEE Trans. Wirel. Commun.* **2014**, *13*, 592–603.
18. Ometov, A.; Olshannikova, E.; Masek, P.; Olsson, T.; Hosek, J.; Andreev, S.; Koucheryavy, Y. Dynamic Trust Associations Over Socially-Aware D2D Technology: A Practical Implementation Perspective. *IEEE Access* **2016**, *4*, 7692–7702.
19. Wu, D.; Zhou, L.; Cai, Y. Social-Aware Rate Based Content Sharing Mode Selection for D2D Content Sharing Scenarios. *IEEE Trans. Multimed.* **2017**, *99*, doi:10.1109/TMM.2017.2700621.
20. Datsika, E.; Antonopoulos, A.; Zorba, N.; Verikoukis, C. Green cooperative device-to-device communication: A social-aware perspective. *IEEE Access* **2016**, *4*, 3697–3707.

21. Huang, Z.; Tian, H.; Fan, S.; Xing, Z.; Zhang, X. Social-Aware Resource Allocation for Content Dissemination Networks: An Evolutionary Game Approach. *IEEE Access* **2016**, *5*, 9568–9579.

22. Wang, Z.; Sun, L.; Zhang, M.; Pang, H.; Tian, E.; Zhu, W. Propagation-and mobility-aware d2d social content replication. *IEEE Trans. Mob. Comput.* **2017**, *16*, 1107–1120.

23. Atzori, L.; Iera, A.; Morabito, G.; Nitti, M. The Social Internet of Things (SIoT)—When social networks meet the Internet of Things: Concept, architecture and network characterization. *Comput. Netw.* **2012**, *56*, 3594–3608.

24. Nitti, M.; Murroni, M.; Fadda, M.; Atzori, L. Exploiting Social Internet of Things Features in Cognitive Radio. *IEEE Access* **2016**, *4*, 9204–9212.

25. Nitti, M.; Girau, R.; Atzori, L. Trustworthiness Management in the Social Internet of Things. *IEEE Trans. Knowl. Data Eng.* **2014**, *26*, 1253–1266.

26. 3GPP. *TS 22.368, Service Requirements for Machine-Type Communications (MTC), V13.1.0*; Technical Report; European Telecommunications Standards Institute: Sophia Antipolis Cedex, France, 2014.

27. 3GPP. *TSG RAN Meeting #69, Narrowband IoT*; Technical Report; European Telecommunications Standards Institute: Sophia Antipolis Cedex, France, 2015.

28. Pan, Y.; Pan, C.; Zhu, H.; Ahmed, Q.Z.; Chen, M.; Wang, J. On consideration of content preference and sharing willingness in D2D assisted offloading. *IEEE J. Sel. Areas Commun.* **2017**, *35*, 978–993.

29. Datsika, E.; Antonopoulos, A.; Zorba, N.; Verikoukis, C. Cross-Network Performance Analysis of Network Coding Aided Cooperative Outband D2D Communications. *IEEE Trans. Wirel. Commun.* **2017**, *16*, 3176–3188.

30. Zhou, B.; Hu, H.; Huang, S.Q.; Chen, H.H. Intracluster device-to-device relay algorithm with optimal resource utilization. *IEEE Trans. Veh. Technol.* **2013**, *62*, 2315–2326.

31. Yang, F.; Wang, S.; Li, J.; Liu, Z.; Sun, Q. An overview of Internet of Vehicles. *China Commun.* **2014**, *11*, 1–15.

32. Iturralde, M.; Yahiya, T.; Wei, A.; Beylot, A. Interference mitigation by dynamic self-power control in femtocell scenarios in LTE networks. *IEEE Glob. Commun. Conf.* **2012**, 4810–4815, doi:10.1109/GLOCOM.2012.6503880.

![future internet logo] *future internet*

MDPI

Article

Social-Aware Relay Selection for Cooperative Multicast Device-to-Device Communications

Francesco Chiti , Romano Fantacci and Laura Pierucci *

Department of Information Engineering, University of Florence, 50139 Florence, Italy;
francesco.chiti@unifi.it (F.C.); romano.fantacci@unifi.it (R.F.)
*   Correspondence: laura.pierucci@unifi.it; Tel.: +39-055-275-8626

Received: 29 September 2017; Accepted: 28 November 2017; Published: 4 December 2017

Abstract: The increasing use of social networks such as Facebook, Twitter, and Instagram to share photos, video streaming, and music among friends has generated a huge increase in the amount of data traffic over wireless networks. This social behavior has triggered new communication paradigms such as device-to-device (D2D) and relaying communication schemes, which are both considered as strong drivers for the next fifth-generation (5G) cellular systems. Recently, the social-aware layer and its relationship to and influence on the physical communications layer have gained great attention as emerging focus points. We focus here on the case of relaying communications to pursue the multicast data dissemination to a group of users forming a social community through a relay node, according to the extension of the D2D mode to the case of device-to-many devices. Moreover, in our case, the source selects the device to act as the relay among different users of the multicast group by taking into account both the propagation link conditions and the relay social-trust level with the constraint of minimizing the end-to-end content delivery delay. An optimization procedure is also proposed in order to achieve the best performance. Finally, numerical results are provided to highlight the advantages of considering the impact of social level on the end-to-end delivery delay in the integrated social–physical network in comparison with the classical relay-assisted multicast communications for which the relay social-trust level is not considered.

Keywords: multicast; device-to-device communications; Internet of Things; mobile social networks

1. Introduction

In the incoming fifth-generation (5G) system, device to device (D2D) and relaying communications are envisaged as enablers to face the huge amount of data traffic due to novel and advanced applications and services. In particular, the emerging trend of sharing photos, video, and music among friends requires a large amount of data at a high data rate. In D2D communications, the devices can share the same relevant contents or help the neighbours to deliver data by establishing a direct link without (or with limited) involvement of a base station (BS) or eNodeB. As the D2D communications occur over short distances, they can support a higher data rate with respect to infrastructure communications. Furthermore, D2D communications enhance the spectral efficiency, lighten areas with elevate traffic and improve the user quality of experience.

In the foreseen integrated 5G and Internet of Things (IoT) infrastructures, short-range D2D communications can interconnect heterogeneous devices with lower-energy consumptions guaranteeing proximity services for 5G/IoT networks. The D2D concept allows for a direct connection between D2D pairs of devices, considerably (or fully) reducing the exchange of traffic requests with the BS.

The D2D approach is also useful in the multicast context, where multiple cellular/IoT devices have to receive the same data from the BS. Various devices can self-organize into clusters, and some of these can be selected to act as a relay to help the forwarding of data, particularly to the end-nodes of the networks and the offloading of traffic from the BS.

Relay-assisted communications have been actively studied and are already considered in the standardization process of mobile broadband communication systems, such as in the Third Generation Partnership Program (3GPP) Long-Term Evolution Advanced (LTE-Advanced), IEEE 802.16j and the IEEE 802.16m [1,2] to improve the cell-edge coverage radius and to provide high data rates to the users located in the cell-edge or in coverage holes.

In addition, cooperative multicast relaying can improve data rates as a result of the shorter distance from multicast devices with respect to the direct transmission, for which the worst propagation channel limits the available transmission rate. In the next generation, 5G networks, cognitive relaying and cooperative D2D relaying will continue to be the main players.

In cooperative D2D communications, the common concept is that all the devices can relay to each other, but, for example, the battery charging can often limit data forwarding assistance because of the energy consumption. If D2D communication is under a partial operator control, the user acting as a relay can have a bill reduction, or the user can offer his battery and bandwidth consumption only if he wants to help his friend.

Recently, the increasing demand for social applications, such as Facebook, Twitter, YouTube, Instagram, and so on, has suggested integrating the social behavior on a cooperative D2D/relaying design, in which the D2D users can communicate directly with each other and can exchange content mainly if they are friends. Different social communities, in which each participant has the same interest in content, can be formed, for example, by tracking friends, kin and colleagues that share content frequently by online social networks. Social media networking indeed represents a disruptive paradigm leading the transition to the "Web2.0". The proliferation of popular applications for smartphones points out a constant trend for geo-referenced services with an increased level of integration among different communities, thus making the content dissemination more pervasive and instantaneous.

Generally, the social trust model for cooperative D2D communications is built by two different layers that interact each other:

- The *online social network layer*, which indicates the different levels of relationships among the D2D users.
- The *offline mobile communications network layer*, which determines the wireless connections subject to the channel propagation conditions.

For example, a high level of relationships can exist between two users of the social layer because they belong to the same group, such as family or colleagues, but the same connection cannot exist in the physical layer because there is no proximity or there are bad propagation conditions of the link, as shown in Figure 1.

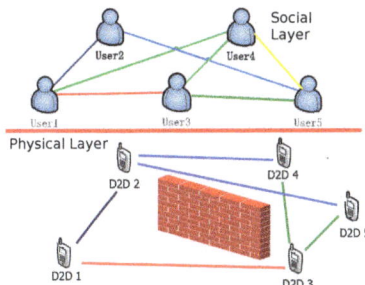

Figure 1. Social–physical layers for cooperative device-to-device (D2D) relaying.

The social-trust level can be evaluated through the analysis of the contents and account information shared, for example, by tracking online social network services and calculating the ranking of the user to spread the content (this is related to the number of friends). The social relationships can be retrieved by matching mobile phones' contact books: if two users are colleagues or members of the same family, it is very likely that they have many phone contacts in common.

Moreover, when the D2D relay has received popular content from the BS, it can share this to other users in proximity through D2D communications, and a new entry in the group of friends can make available this content, considerably decreasing the traffic load of the BS. The geographical location can impact this social transfer of the same data; for example, in a school building, it is highly likely that different students download the same viral video, and thus a proximity area can be a characteristic for content delivery.

Current researches analyze the cooperative communication gain for proper relay selection by considering a weighted tradeoff between the social trust information and the physical constraints in order to maximize the throughput with respect to the direct transmission, mainly for scenarios with only one destination.

This paper presents a multicast scenario with a source (either the eNodeB or a D2D user), which transmits the same data content to many destinations through the selection of the best relay among different alternatives, by taking into account the relay social-trust level and propagation condition links to optimize the end-to-end delivery delay. This is defined as the time required to deliver data from the source through the relay to all the destinations belonging to the multicast group (i.e., the social community). The relay itself is interested in receiving data content because it can belong to a multicast group. The social-trust level is related to the social relationships of the relay with the source and represents the part of the transmit power that the relay is available to give *as friend* to forward the data to the multicast community and as a consequence, the availability to consume its own battery. The performance of this social cooperative multicast system is analyzed by considering that the direct link among the source and destinations is not available because of path loss and shadowing effects and that all the D2D multicast devices have to receive all the data.

The remainder of the paper is organized as follows. Section 2 provides a literature survey. In Section 3, the system model is introduced, and in Section 4, we provide the simulation results for the end-to-end delivery time performance evaluation of the integrated social multicast D2D-based system. Finally, the conclusions are drawn.

2. Literature Review

D2D communications are foreseen to be of paramount importance in 5G systems for improving system capacity and for offloading traffic from a BS. In this context, devices can autonomously establish direct connections sharing the spectrum of cellular systems (underlaid D2D communications) and resource allocation, and management approaches have to be handled to provide proximity services [3,4]. Different research proposes interference-avoiding schemes under the management of network infrastructure to prevent harmful interference among D2D and cellular users or to analyze autonomous D2D data transmissions with guaranteed limited and tolerable interference on cellular users. Others challenges faced regard the significant reduction of traffic load and communication delay [5] and the neighbor discovery methods to detect proximity users in cellular networks with underlaid D2D communications [6].

As users can self-organize with direct connections, D2D cooperation can be the main means to improve throughput and energy efficiency. Unfortunately, when D2D users cooperate, the devices acting as relays expend extraordinary energy for data transmissions, even over short distances. Hence, it is necessary to select D2D relays among multiple devices by considering whether a device has already been selected as a relay many times or if it has to handle too many cooperative users to avoid excessive energy consumption of the relay and to decrease the life-time of the system [7].

In D2D communications networks, the mobile devices are intermittently connected in an ad hoc manner, and the topology of the network can be highly dynamic. In [8], a novel opportunistic network routing protocol based on social rank and intermeeting time is considered. Cooperative multicast transmissions, for which different devices with good channel conditions are selected as relay nodes, improve the achievable data rate with respect to the direct transmission from the BS, which suffers from the constraint of the worst propagation channel as a result of the long distance, for example. In the literature, various relaying technologies are considered, such as decode and forward (DF) and amplify-and-forward (AM) relaying. Different relaying strategies have been proposed, such as relay selection, in which multiple relay nodes allow for a more efficient use of the system resources through the selection of the best source–relay and relay–destination channel, and incremental relaying, in which feedback from the destination about the success or failure of direct source–destination transmission is used. Full duplex (FD) relays offer high spectral efficiency but suffer from strong self-interference and loop interference if multiple antennas are installed on each relay. On the other hand, half-duplex (HD) relays, for which the source transmits information to relays in the broadcast phase and then relays this forward to destinations in the successive time slots, causes multiplexing loss. Many methods have recently been proposed to overcome the multiplexing loss in HD relaying. Successive relay techniques [9] are analyzed to improve the spectral efficiency of HD relays, in which a pair of relays is selected, and while one relay receives data from the source, the other transmits the previously received data to the destination. The use of a buffer at the relay nodes, "buffer-aided relaying methods", as in [10–15], increases the spectral efficiency with respect to HD relaying without a buffer. This approach allows for selecting the best HD buffer-aided relay among the various relay–destination links (opportunistic relaying schemes), to transmit the data if the channel conditions are good or buffer it otherwise, as in max-max relay selection by Ikhlef et al. [16] or the max-link relay selection scheme by Krikidis et al. [17].

Recently, several analyses of user behaviors and their social relationships are addressed by considering the most popular content shared on social networks such as Facebook, Instagram, YouTube, and WhatsApp, as well what the influences of other friends, media, bloggers and advertising are. Different probabilistic models are proposed to predict the rate of downloading and the requests' evolution of some content.

The new vision is to join the social relationships information and proximity-based communications to build an autonomous, trustworthy network of smart IoT devices. The impact of social relationships on the overall throughput of a D2D communication system is considered in [18] with the stop–wait approach for relay selection. The distributed resource allocation, mainly based on cooperative game theory to exploit diverse social relationships on a physical domain, is analyzed in [19]; a Bayesian model for social relationships and a coalitional graph game for efficient data distribution is proposed in [20]; the social characteristics are used to help ad hoc peer discovery in [21]; a neighbor discovery method and a dynamic detection of overlapping social communities is highlighted in [6]. Zhang et al. [22] propose a model for the delivery of content in the online social level jointly with the optimization of a traffic offloading process for the D2D communications layer. In [23], the sociality among IoT devices is used to model the trustworthiness for successful D2D-based content delivery to significantly reduce the impact of malicious behavior.

In [18], the optimal stopping policy for relay selection is proposed for the case of cooperative D2D relaying, but the multicast context is not considered. This policy, suitably elaborated for the multicast case, is considered as the benchmark for our throughput performance. In [24], a cooperative multicast scheme with underlaid D2D communications is proposed, in which social relationships drive the relay selection. The selected relay nodes receive broadcast messages from the BS on downlink (DL) band and then forward these to multicast users in the uplink (UL) band via D2D communications to increase the multicast transmission rate.

In the proposed paper, a cooperative multicast scenario is also considered as in [24] with the difference that the best relay is selected to optimize the global end-to-end delivery delay metric.

Moreover, the same content broadcast from the BS to the relay is distributed to the users of the multicast group by assuming that one message is delivered only when all the users in the multicast group have received it. We would like to stress that our focus is on a social–physical scenario, for which, in particular, the social level trust is considered to have a more general meaning with respect to classical real-world networks. In particular, the social level trust of a device is related to its need or interest in receiving a given information flow, for example, according to the novel paradigm of Fog computing/networking and applications in which clusters of smart devices collaborate toward a common goal. As a consequence, the end-to-end delivery delay is considered as the most important parameter to evaluate the optimal social-aware relay selection mechanism in the case of multicast transmission. Therefore, in this paper, the best relay is selected among the different D2D devices according to the sociality index, the proximity distance to the multicast users and the links' propagation conditions.

3. System Model

We focus on a cooperative multicast context, in which end users receive the requested content not directly from the BS but via other users acting as a relay in D2D communications.

In the considered social–physical scenario, the users can communicate directly with other multicast users and exchange shared content mainly if they are friends and according to the quality of the D2D connection links. In particular, we refer here to the classical social–physical architecture, which entails two layers, as in Figure 1. In this context, the source (either the eNodeB or a D2D node) can deliver the same content to a group of D2D devices in a small area, and as the node density increases, various devices can act as relays to forward multicast traffic to the end-devices of the social communities and vice versa. The devices selected as relays may also be interested in receiving the same data content because they could belong to the multicast group. According to the multicast concept, data is delivered if all the destinations have received it. Moreover, we assume that the D2D multicast group can be very far from the BS and that the direct path cannot exist or suffers from a deep attenuation due to path-loss and shadowing effects.

A device can leave or join the multicast community according to his mobility. However, we assume that the D2D devices remain in the same location during a D2D communication transmission period (e.g., in the order of milliseconds), while their positions can change across different periods because of users' mobility. The D2D social-trust levels and the forwarding metric related to the channel quality are known at the source and D2D/relays.

We consider a relay-assisted network in which a source S transmits to many destinations D_i with $i = 1....M$ through different relays R_j with $j = 1, ...N$, as in Figure 2. We assume flat block fading on all the links, such that the channel coefficients can be assumed to be constant during one time transmission period and can change from one period to the next. One relay is selected among N to serve the multiple destinations according to the best joint quality of the social and physical layers. The relay nodes use HD decoding and the forward relaying mode. The nodes can retrieve their social relationships information, for example, from the BS, reporting the analysis of content sharing or account information by accessing online social websites such as Facebook, Twitter, YouTube, Instagram, and so forth. These social trust values tend to be stable over the time reporting friendships and acquaintances among users. Moreover, for example, Twitter associates with each tweet the exact location (latitude and longitude), and this might allow the nodes the possibility to know their related positions and also discover their neighbors in mobility.

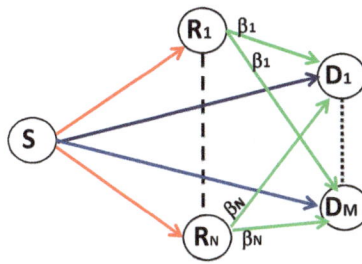

Figure 2. Multicast model based on social-trust level.

As in [18], the social trust impacts the transmission power of the relay node: if a stronger social-trust relationship exists between the source and relay, the relay provides more power resources to help the source node in retransmission.

Therefore, the selected relay R_j transmits to the multiple destinations with a power proportional to the strength of the social-trust value β_j. For each relay–destination link, the γ signal-to-noise ratio (SNR) at the destination D_i is related to the path-loss, the Rayleigh flat fading, and the zero-mean additive white Gaussian noise (AWGN) with variance σ_n^2, and it can be expressed as

$$\gamma_{R_j,D_i} = \frac{P_{R_j}|h_{R_j,D_i}|^2}{\sigma_n^2 d_{R_j,D_j}^\alpha} \tag{1}$$

where P_{R_j} is the transmit power of the jth relay, d_{R_j,D_i} stands for the distance of relay R_j from destinations D_i, α is the path-loss coefficient, and h_{R_j,D_i} are the Rayleigh fading coefficients.

In the multicast transmission mode, minimizing the delivery latency necessary to receive the data to all the destinations is the main objective.

As a consequence, the aim of this paper is to minimize the end-to-end delivery delay that is obtained considering T_j, the time needed to transmit from the source to the jth selected relay, and $T_{j,I}$, the overall time for transmission from the relay to the destinations. In the case of multicast transmission, the time to deliver the data from the relay to destinations is related not only to the link with the higher delay, because all the destinations have to receive the same content, but also to the social-trust level of the relay, which allows or does not allow the transmission to destinations. The quality of physical channels is directly related to the SNR and consequently to the data rate available, while the social trust β_j assures the social link among the nodes. These two aspects, the social relationship and the channel quality, must to be balanced to optimize the end-to-end delivery delay for the cooperative multicast D2D relaying.

Accordingly to this, the channel rate C_j for the link between the source and the jth relay is

$$C_j = W \log_2(1 + \gamma_{S,R_j}) \tag{2}$$

where W is the system bandwidth and γ_{S,R_j} is the SNR of the link between the source and the jth relay; the multicast channel rate can be expressed for each link from the jth relay to all the destinations D_i with $i = 1...M$ as

$$C_{j,i} = W\log_2(1 + \gamma_{R_j,D_i}\beta_j) \tag{3}$$

The data rate for cooperative D2D relaying is

$$C_{j,I} = \min_i(C_{j,i}) \tag{4}$$

Therefore, by considering the impact of the social trust β_j, the delivery delays are

$$T_j \simeq \frac{1}{C_j} \tag{5}$$

$$T_{j,I} \simeq \frac{1}{C_{j,I}} \tag{6}$$

Then, $T_{j,I}$ is the maximum delivery delay related to the worst link to the relay destinations. We consider that the transmit power that needs to transmit in two hops, source–relay and relay–destinations, is not larger than that of the multicast transmission without the use of relays. Regarding the social-trust level β_j, we assume two values, 0 and 1, where a higher value of β_j represents a stronger social trust; moreover, we consider that friends have similar behavior in transmitting shared data. For example, a D2D user/relay can choose its friends on its own contact list to define who and how it can help other users for cooperative D2D relaying.

To this end, the optimization problem can be formulated and approximated as

$$\min_j (T_j + T_{j,I}) \tag{7}$$

$$s.t. \ T_j, T_{j,I} > 0 \tag{8}$$

$$P_S T_j + P_{R_j} T_{j,I} < P_{S_1} T_1 \tag{9}$$

The second constraint (Equation (9)) guarantees that the total transmit power for the two hops due to relay transmission is not larger than that of multicast communication, for which the source transmits with a power P_{S_1} directly to the destinations over a total time duration equal to T_1. The power P_S transmitted by the source for the two hops can be different from the power P_{S_1}.

The optimization problem (Equation (7)) can be solved with a complexity of $O(M * N)$. An exhaustive search to find the optimal solution is affordable, because in the case of multicast communications, the number of participants to the multicast cluster, and consequently the subset of relays, is limited.

In the social relay selection method, we first retrieve the payable transmit power, the geographical locations of each device, and the channel gains and calculate the SNR for each link. Then, we calculate the delivery time needed from the source to each relay and the maximum delivery delay for each group of relay destinations sequentially until the number of relays is reached. We sort the sum related to these delivery times, and, finally, we select the best relay, which assures the minimum delivery time from the source to the destinations. These steps are shown in Algorithm 1 following.

Algorithm 1: Social Relays Selection Procedure

Input : Number of relays N, Number of Destinations M, Physical Network distances and Rayleigh

coefficients, Social network levels β_j

1 **for** $j \in N$ **do**

2   Calculate $\gamma_{S,R_j} = \frac{P_S |h_{S,R_j}|^2}{\sigma_n^2 d_{S,R_j}^\alpha}$

3   Calculate $C_j = W \log_2(1 + \gamma_{S,R_j})$

4   Calculate $T_j = \frac{1}{C_j}$

5   **for** $i \in M$ **do**

6   Calculate $\gamma_{R_j,D_i} = \frac{P_{R_j} |h_{R_j,D_i}|^2}{\sigma_n^2 d_{R_j,D_j}^\alpha}$

7   Calculate $C_{j,i} = W log_2(1 + \gamma_{R_j,D_i} \beta_j)$

8   **end**

9   Calculate $T_{j,I} = \frac{1}{min(C_{j,i})}$

10 **end**

11 $sort(T + T_I)$ a vector of N elements

12 **while** $k \in N$ **do**

13   Find $(T_k + T_{k,I})$

14   **if** $P_S T + P_R T_I < P_{S_1} T_1$ **then**

15   **return** k optimal solution of (7) and stop

16   **end**

17 **end**

18 Update the new values of β_j and SNR for each links for searching the best relay in the next transmission

period

---

4. Numerical Results

In this section, we provide numerical results concerning the performance of the proposed D2D multicast social relaying method. In the considered scenario, a single D2D source device with R_j neighbor relay devices, where $j = 1...N$, and D_i destination devices, where $i = 1...M$, are located in a single cell with radius C, which guarantees a reserved channel for the control communications among these devices. Destination devices are distributed in a restricted area to form a cluster with radius r and $0 \leq r \leq C$. Within the D2D multicast area, the *i*th device is at the location (d_i, θ_i) with $-r \leq d_i \leq r$ and $0 \leq \theta_i \leq 2\pi$, and the distance between two nodes is defined as $d_{j,i} = \sqrt{d_j^2 + d_i^2 - 2d_j d_i cos(\theta_j - \theta_i)}$. We assume that the distance between the source and the destinations is larger than the distance between any two devices in the multicast group and among the relay and devices and is heavily attenuated because of obstacles, building, and so on. Therefore, the direct link between the source and the destinations does not exist, and communications can be established only via relay.

We consider a transmission channel model often assumed in literature [17], including both the large scale path loss, shadowing variations and zero-mean AWGN. We assume the frequency of non-selective Rayleigh block fading according to a complex Gaussian distribution with zero mean and variance $\sigma_{j,i}^2$ for the *i*th to *j*th link, that is, constant during one transmission period and changing independently from one period to another. The channel gains $|h_{j,i}|^2$ are exponentially distributed, as in [10,17].

The average SNR for the signal received by a generic D2D destination device from the D2D source device is defined as γ_{SD}, and the standard deviation is defined as σ_{sd}^2. To simulate an urban environment, we set the γ_{SD} value in the region of 0 dB. This choice considers that the direct path has a heavier attenuation, as a result, for example, of obstacles, buildings, and so on, in a generic real multicast D2D communication scenario. The average SNR for the signal received by a D2D destination device from one of the D2D relays is defined as γ_{RD}, and the average SNR for the signal received by a generic jth D2D relay from the D2D source device is γ_{SR}.

Table 1 shows the values of the main parameters. The value of β_j, as described in the system model, varies depending of the friendship level. If the D2D relay has a friendship connection, it sets its β value equal to 1; instead, if it has no friendship connection, it reduces the value 10-fold. We suppose that each relay and the D2D source device have the complete knowledge of friendship relations and channel state in terms of the SNR.

Table 1. Main Simulation Parameters.

Bandwidth	10 MHz
Path-loss coefficient	4
Cell radius	1 Km
Cluster radius	0.03 Km
Number of destinations, M	10
Number of relays, N	5:15
γ_{SD}	3 dB
γ_{RD}	13–18 dB
Friendship probability	20%
β_j for a friend source	1
β_j for a non-friend source	0.1
σ_{sd}	1 dB
σ_{rd}	9 dB
User distribution	Uniform

For a comparative perspective, the throughput normalized with respect to the throughput of direct source–destination transmissions, that is, without the use of relays, is shown in Figure 3 for the number of relays increasing. First, a comparison of the normalized throughput of our method with social and physical information with respect to the stopping approach [18] is shown in the top of Figure 3, for which the direct transmission is added to the source–relay transmission at the receiver, highlighting similar performance.

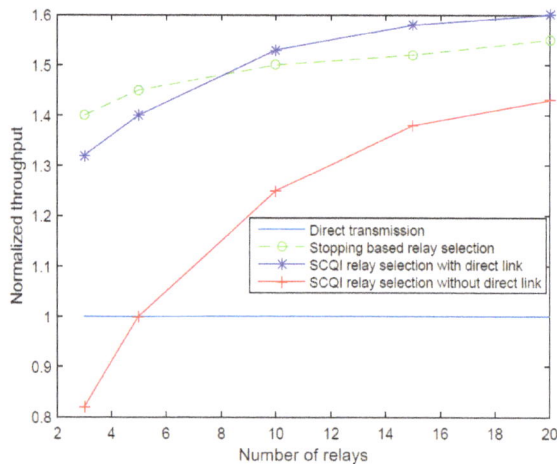

Figure 3. Comparison of normalized throughput with respect to the number of relays for the cases with and without direct transmission.

If obstacles impact heavily on the transmission, that is, the non-line-of-sight (NLOS) case, the performance degrades, as in Figure 3; this is the more interesting case to analyze because it is closer to real environments. Therefore, in the following, the results refer to the case without direct transmission for the parameter mainly related to the multicast transmission, that is, the end-to-end delivery delay. To emphasize the relevance of social relations in Figure 4, we compare our method, the Social Channel Quality Indicator (SCQI), with two alternatives. In the first, the source selects the relay by taking into account only the friendship level. In particular, it randomly selects a relay from the friends without considering the other neighbor relays. In the second comparison method, the D2D source selects the relay with the minimum delay among the source–relays–destination links only relying on the SNR report. Figure 4 shows the average end-to-end delivery delay for a multicast transmission among the source and all the destinations through the single relay selected, introduced by these three alternatives. To simplify the analysis of the results, the average delays are normalized with respect to the direct-link communication delay obtained for the case that the D2D source can communicate directly with the destinations. Figure 4 shows the importance of the knowledge of both friendship and channel conditions. Our method, which considers both of these, guarantees better performance in terms of transmission delays. Again, the direct link is only shown for reference in this figure.

To evaluate the performance of our method relative to the probability for the relay to be a friend with the source in terms of delivery delay, we consider different friendship probability values, maintaining the same system parameters and channel conditions. The results in Figure 5 show that if the total number of neighbor relays grows, the influence of friendship probability is reduced. Each relay addition improves the delivery delay, as we have a greater probability to find friends that offer their full support in retransmission with the lowest latency in the two-way multicasting.

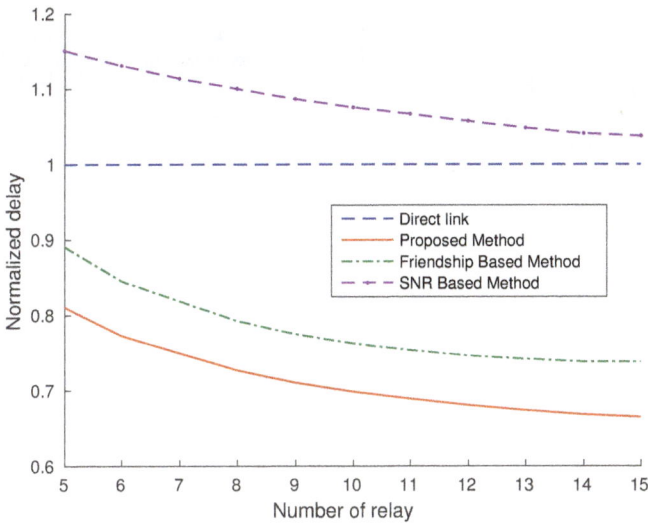

Figure 4. Normalized delay with respect to the number of relays for the considered methods; $\gamma_{RD} = 13$ dB, $\gamma_{SD} = 3$ dB, $\sigma_{sd} = 1$ dB, and $\sigma_{rd} = 2$ dB.

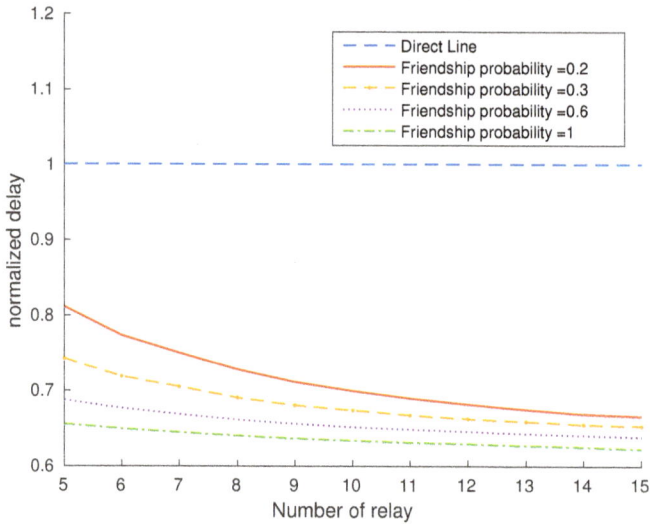

Figure 5. Normalized delay with respect to the number of relays varying the friendship probability; $\gamma_{RD} = 13$ dB, $\gamma_{SD} = 3$ dB, and $\sigma_{sd} = 1$ dB.

Finally, Figure 6 shows the performance of our method as a function of different γ_{RD} values. From this figure, we can note a trend: the normalized delay increases as the SNR_{rd} value decreases and approaches the value of the direct link depending on the number of relays. For example, in the case of γ_{RD} equal to 13 dB, we can note that the direct link is convenient for a number of relays of about seven. This trend is present in each relay-based model: when γ_{SR}, γ_{RD} and γ_{SD} have similar values, the direct communication between the source and destinations becomes the best solution.

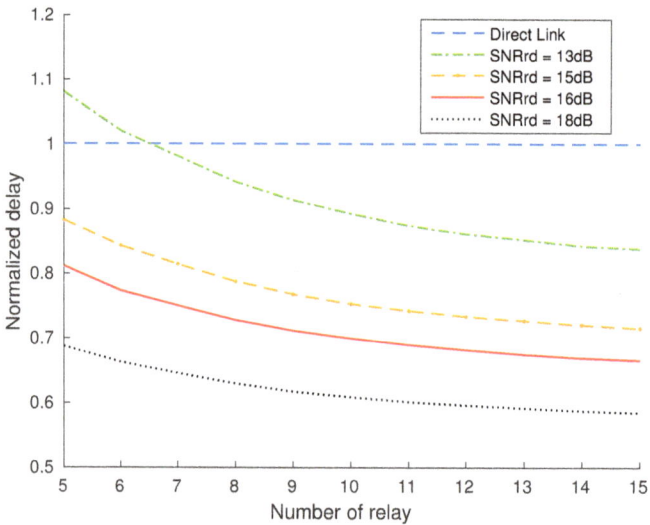

Figure 6. Normalized delay with respect to the number of relays for different γ_{RD} values; $\gamma_{SD} = 3$ dB, and $\sigma_{sd} = 1$ dB.

5. Conclusions

In this paper, we propose a relay selection method for D2D multicast communication based on D2D channel conditions and social-trust levels. We derive the optimal social-aware relay selection method on the basis of the minimization of the multicast end-to-end delivery time. We further show that our method for social-aware relay selection exhibits an incremental performance with respect to the method that does not use the social domain information, and vice versa for the method in which the relay selection is only based on friendship informations. Numerical results demonstrate that the proposed mechanism can achieve better performance gain when the difference between the channel quality in the two-way source–relay–destinations is better with respect to the source–destinations channel condition.

We have also highlighted the impact of friendship in the decision method. A larger number of friends in the neighborhood achieves a better average performance in terms of transmission delay. Our proposed method can be used in a cell where all the multicast participants are distributed within a small area to reduce the amount of network traffic necessary to deliver digital content to all users belonging to the same social community.

Future developments can be represented by the extension of this method to multi-hop social-aware content distribution, in which each destination is interested in sharing the same digital content with a different community, for example, according to the novel paradigm of fog computing/networking and applications in which clusters of smart devices collaborate toward a common goal.

Furthermore, we can use real-world human mobility traces such as Intel, Infocom06 or Brightkite [25] for future simulations by dividing the devices in the network into several groups to evaluate the system performance in terms of end-to-end delivery delay in realistic environments.

Acknowledgments: This work was partially supported by the project "GAUChO" funded by Progetti di Ricerca di Elevato Interesse Nazionale (PRIN), 2015. The authors thank Dario Di Giacomo for his work to optimize the algorithm.

Author Contributions: All the authors contributed equally to the paper. R. Fantacci coordinated the research activities and supervised the work; F. Chiti and L. Pierucci mainly worked on the system model, designed the experiments and analyzed the simulation results. All the authors wrote the paper and reviewed its structure and intellectual content.

Conflicts of Interest: The authors declare no conflict of interest.

References

1. Yang, Y.; Hu, H.; Xu, J.; Mao, G. Relay technologies for WiMax and LTE-advanced mobile systems. *IEEE Commun. Mag.* **2009**, *47*, 100–105.
2. Nomikos, N.; Skoutas, D.; Makris, P. Relay selection in 5G networks. In Proceedings of the 2014 International Wireless Communications and Mobile Computing Conference (IWCMC), Nicosia, Cyprus, 4–8 August 2014; pp. 821–826.
3. Yu, C.H.; Doppler, K.; Ribeiro, C.B.; Tirkkonen, O. Resource Sharing Optimization for Device-to-Device Communication Underlaying Cellular Networks. *IEEE Trans. Wirel. Commun.* **2011**, *10*, 2752–2763.
4. Chiti, F.; Giacomo, D.D.; Fantacci, R.; Pierucci, L. Interference aware approach for D2D communications. In Proceedings of the 2016 IEEE International Conference on Communications (ICC), Kuala Lumpur, Malaysia, 22–27 May 2016; pp. 1–6.
5. Koskela, T.; Hakola, S.; Chen, T.; Lehtomaki, J. Clustering Concept Using Device-To-Device Communication in Cellular System. In Proceedings of the 2010 IEEE Wireless Communication and Networking Conference, Sydney, Australia, 18–21 April 2010; pp. 1–6.
6. Wang, R.; Yang, H.; Wang, H.; Wu, D. Social overlapping community-aware neighbor discovery for D2D communications. *IEEE Wirel. Commun.* **2016**, *23*, 28–34.
7. Zhang, Z.; Wang, L.; Zhang, J. Energy Efficiency of D2D Multi-User Cooperation. *Sensors* **2017**, doi:10.3390/s17040697.

8. Wang, T.; Zhou, Y.; Wang, Y.; Tang, M. Novel Opportunistic Network Routing Based on Social Rank for Device-to-Device Communication. *J. Comput. Netw. Commun.* **2017**, *2017*, 2717403.
9. Tannious, R.; Nosratinia, A. Spectrally-efficient relay selection with limited feedback. *IEEE J. Sel. Areas Commun.* **2008**, *26*, 1419–1428.
10. Zlatanov, N.; Schober, R. Buffer-Aided Relaying With Adaptive Link Selection-Fixed and Mixed Rate Transmission. *IEEE Trans. Inf. Theory* **2013**, *59*, 2816–2840.
11. Simoni, R.; Jamali, V.; Zlatanov, N.; Schober, R.; Pierucci, L.; Fantacci, R. Buffer-Aided Diamond Relay Network with Block Fading. In Proceedings of the IEEE International conference on communications (ICC2015), London, UK, 8–12 June 2015.
12. Simoni, R.; Jamali, V.; Zlatanov, N.; Schober, R.; Pierucci, L.; Fantacci, R. Buffer-Aided Diamond Relay Network With Block Fading and Inter-Relay Interference. *IEEE Trans. Wirel. Commun.* **2016**, *15*, 7357–7372.
13. Chiti, F.; Fantacci, R.; Pierucci, L. Buffer-aided relaying approaches for multicast communications. In Proceedings of the 2016 International Wireless Communications and Mobile Computing Conference (IWCMC), Paphos, Cyprus, 5–9 September 2016; pp. 411–416.
14. Chiti, F.; Fantacci, R.; Pierucci, L. Dynamic multicast link selections for buffer-aided relaying networks. *Int. J. Commun. Syst.* **2016**, *29*, 1790–1804.
15. Chiti, F.; Fantacci, R.; Pierucci, L.; Privitera, N. Optimal joint MIMO and modulation order selection for network coded multicast wireless communications. *Telecommun. Syst.* **2015**, *61*, 1–9.
16. Ikhlef, A.; Michalopoulos, D.; Schober, R. Max-Max Relay Selection for Relays with Buffers. *IEEE Trans. Wirel. Commun.* **2012**, *11*, 1124–1135.
17. Krikidis, I.; Charalambous, T.; Thompson, J. Buffer-Aided Relay Selection for Cooperative Diversity Systems without Delay Constraints. *IEEE Trans. Wirel. Commun.* **2012**, *11*, 1957–1967.
18. Zhang, M.; Chen, X.; Zhang, J. Social-aware relay selection for cooperative networking: An optimal stopping approach. In Proceedings of the 2014 IEEE International Conference on Communications (ICC), Sydney, Australia, 10–14 June 2014; pp. 2257–2262.
19. Zhao, Y.; Li, Y.; Cao, Y.; Jiang, T.; Ge, N. Social-Aware Resource Allocation for Device-to-Device Communications Underlaying Cellular Networks. *IEEE Trans. Wirel. Commun.* **2015**, *14*, 6621–6634.
20. Sun, Y.; Wang, T.; Song, L.; Han, Z. Efficient resource allocation for mobile social networks in D2D communication underlaying cellular networks. In Proceedings of the 2014 IEEE International Conference on Communications (ICC), Sydney, Australia, 10–14 June 2014; pp. 2466–2471.
21. Zhang, B.; Li, Y.; Jin, D.; Hui, P.; Han, Z. Social-Aware Peer Discovery for D2D Communications Underlaying Cellular Networks. *IEEE Trans. Wirel. Commun.* **2015**, *14*, 2426–2439.
22. Zhang, Y.; Pan, E.; Song, L.; Saad, W.; Dawy, Z.; Han, Z. Social Network Aware Device-to-Device Communication in Wireless Networks. *IEEE Trans. Wirel. Commun.* **2015**, *14*, 177–190.
23. Militano, L.; Orsino, A.; Araniti, G.; Iera, A. NB-IoT for D2D-Enhanced Content Uploading with Social Trustworthiness in 5G Systems. *Future Internet* **2017**, *9*, doi:10.3390/fi9030031.
24. Xu, W.; Li, S.; Xu, Y.; Lin, X. Underlaid-D2D-assisted cooperative multicast based on social networks. *Peer-to-Peer Netw. Appl.* **2016**, *9*, 923–935.
25. Cho, E.; Myers, S.A.; Leskovec, J. Friendship and Mobility: Friendship and Mobility: User Movement in Location-Based Social Networks. In Proceedings of the 17th ACM SIGKDD International Conference on Knowledge Discovery and Data Mining, San Diego, CA, USA, 21–24 August 2011.

![future internet logo] *future internet*

MDPI

Conference Report

A Fast and Reliable Broadcast Service for LTE-Advanced Exploiting Multihop Device-to-Device Transmissions

Giovanni Nardini , Giovanni Stea * and Antonio Virdis

Dipartimento di Ingegneria dell'Informazione, University of Pisa, Largo Lucio Lazzarino 1, 56122 Pisa, Italy;
g.nardini@ing.unipi.it (G.N.); a.virdis@iet.unipi.it (A.V.)
* Correspondence: giovanni.stea@unipi.it; Tel.: +39-050-2217653

Received: 13 October 2017; Accepted: 21 November 2017; Published: 25 November 2017

Abstract: Several applications, from the Internet of Things for smart cities to those for vehicular networks, need fast and reliable proximity-based broadcast communications, i.e., the ability to reach all peers in a geographical neighborhood around the originator of a message, as well as ubiquitous connectivity. In this paper, we point out the inherent limitations of the LTE (Long-Term Evolution) cellular network, which make it difficult, if possible at all, to engineer such a service using traditional infrastructure-based communications. We argue, instead, that network-controlled device-to-device (D2D) communications, relayed in a multihop fashion, can efficiently support this service. To substantiate the above claim, we design a proximity-based broadcast service which exploits multihop D2D. We discuss the relevant issues both at the UE (User Equipment), which has to run applications, and within the network (i.e., at the eNodeBs), where suitable resource allocation schemes have to be enforced. We evaluate the performance of a multihop D2D broadcasting using system-level simulations, and demonstrate that it is fast, reliable and economical from a resource consumption standpoint.

Keywords: proximity services; device-to-device; multihop; resource allocation; mobile networks; vehicular networks; simulation

1. Introduction

The diffusion of sensors and personal devices has recently made possible a range of networked applications that have geographical proximity as a key characteristic. Relevant examples abound: smart-city applications are often based on querying sensors deployed in a certain area (e.g., for temperature, air pollution, etc.) [1]. In vehicle-to-vehicle (V2V) communications, cars that sense anomalous conditions (e.g., a collision) should broadcast this information to their neighbors, to instruct their assisted-driving systems to activate safety maneuvers [2]. Likewise, coordinated robots or drones need to broadcast their position and status to their neighbors to coordinate swarming [3]. In all the above cases, the set of potentially interested recipients of a message generated by an application is defined according to geographical proximity to the originator: anyone close enough should pay attention to the message, where how close is close enough is actually determined by the application itself. For instance, still in the case of vehicular collision, it is foreseeable that only cars in a small radius from the collision point should activate their assisted-driving system and initiate safety maneuvers, whereas vehicle navigation systems in a much larger radius may benefit from knowing about the collision and start looking for alternative routes. In other words, the broadcast domain should be defined directly by the application.

All the above applications need to rely on ubiquitous and reliable and secure connectivity, as well as mobility support. Another requirement of these applications is small latency, either because of a specific deadline, or because the performance of networked applications relying on these broadcast messages depends on how fast these propagate.

In the last decade, researchers and manufacturers widely investigated the performance of 802.11p as a technology for vehicular mobile ad hoc networks. On one hand, the latter has proven to be very scalable and flexible, as it does not need any infrastructure to work. On the other hand, 802.11p has demonstrated to have limitations in providing bounded delay and QoS (Quality of Service) guarantees [4]. Recently, both researchers and automotive industries have begun to investigate using 4G cellular networks, such as LTE-A (Long-Term Evolution-Advanced) as an option for vehicular communications [5]. Research works have evaluated the performance of 4G for various vehicular applications, showing that it can be considered a viable alternative to 802.11p [6]. The above considerations also fall into the context of Vehicle-to-Everything (V2X) communications, where one endpoint of the communications is a vehicle, and the other one can be user cell phones, connected traffic lights, etc. Moreover, several of the above examples of applications are being mentioned as use-cases to generate requirements for the definition of the future 5G communications [7]. We are thus moving towards a context where cellular communications are expected to play a major role as a unifying technology for multiple services.

The current LTE-Advanced standard, unfortunately, is ill equipped to support this type of applications. In fact, cellular communications normally have the eNodeB (eNB) as an endpoint of each layer-2 radio transmission. This requires the User Equipment (UE) application originating the message to always use the eNB as a relay in a two-hop path, even though the destination is a proximate UE. The eNB can relay the message using either multicast or unicast downlink transmissions. The multicast leverages the standard Multicast/Broadcast SubFrame Network, (MBSFN), which was designed for broadcast services like TV. MBSFN is inflexible for at least three reasons: first, multicast/broadcast subframes are alternative to unicast ones, and their definition must be configured semi-statically. Thus, defining MBSFN subframes implies eating into the capacity for downlink unicast transmissions, and reserving capacity for broadcast ones even when there is nothing to relay. If the network is configured to have just one MBSFN subframe per frame (a frame being 10 subframes), then unicast transmission capacity in the downlink is reduced by 10%, and the worst-case delay for a multicast relaying is still 10 ms, which is non negligible. Clearly, this mechanism is tailored to a continuous, periodic traffic, rather than a sporadic, infrequent one. Second, MBSFN transmissions reach a tracking area, which corresponds to a set of cells. There is no way to geofence the broadcast to smaller, user-defined areas. Third, a single transmission format is selected for the whole tracking area: depending on their channel conditions, some—possibly many—UEs may not be able to decode the message, hence reliable delivery is not guaranteed.

eNB-driven relaying using unicast transmissions solve all the above three problems: assuming that the eNB possesses the location of the target UEs (something which is achievable through localization services, empowered by Mobile-edge Computing (MEC) [8]), the eNB may select which UEs to target (hence defining its own geofence), use different transmission formats in order to match their channel conditions, and allocate capacity only on demand. The downside, however, is that this may be too costly in terms of downlink resources: in fact, the resource occupancy grows linearly with the number of UEs. If a 40-byte message has to be relayed to 100 UEs, and the average Channel Quality Indicator (CQI) is 5, three Resource Blocks (RBs) per UE are needed, which means 300 RBs in total, i.e., six subframes entirely devoted to relaying the message within the cell in a 10-MHz bandwidth LTE deployment. This deprives other UEs of bandwidth for a non-negligible time, and consumes energy in the network.

Starting from the latest releases, the LTE-A standard has incorporated network-controlled device-to-device (D2D) transmissions, i.e., broadcast transmissions where both endpoints are UEs. These are also foreseen in the upcoming 5G standard. The eNB still allocates the resources for D2D transmission on the so-called sidelink (SL), which is often physically allocated in the UL (uplink) frame [9]. D2D transmissions have a number of attractive features: they do not increase the operator's energy bill, since data-plane transmissions do not involve the eNB. Moreover, they can occur at reduced power, hence exploit spatial frequency reuse. However, the main downside is that their coverage area is limited to a UE's transmission radius, which is often too small.

This paper, which extends our previous work [10], advocates using multihop D2D transmissions to support geographically constrained broadcast services. Multihopping allows these services to scale up to a larger geographical reach, while retaining all the benefits of D2D. In order to engineer a Multihop D2D-based broadcast (MDB) service, it is necessary to enlist the cooperation of both the UE and the eNB, which are interdependent. In fact, UEs must define the broadcast domain, and—being the only nodes that can use the SL—decide if and when to relay messages, keeping into account that an aggressive relaying policy may waste resources or even induce collisions. On the other hand, the network—and, specifically, the eNBs—must allocate resources to allow the diffusion of the broadcast, and possibly coordinating with neighboring eNBs. Resource allocation can be either static or dynamic (i.e., on demand) [11], and both solutions have pros and cons. Our goal is to prove that an MDB service can be realized by using minimal, standard-compliant cooperation from the network infrastructure (which need not even be aware of the very existence of the MDB service), and only running relatively simple application logic within UEs. While several other papers have investigated multihop D2D transmissions in LTE-A (e.g., [12–17]), this work and its predecessor [10] have been the first paper to propose multihop D2D as a building block for geofenced broadcast services. A relevant issue, therefore, is to investigate what performance can be expected from such services, i.e., what latency, resource consumption, and reliability (i.e., percentage of reached destinations within the broadcast domain) are in order. This paper extends [10] by presenting a thorough discussion and evaluation of the various factors that determine the performance of MDB. Moreover, we discuss the impact on the performance of MDB of different network conditions, such as varying UE density, presence of selfish users, or the occurrence of near-simultaneous broadcasts related to the same event. Last, but not least, we assess the performance of a real-life service, i.e., the diffusion of alerts in a vehicular network scenario, run on MDB. Our results confirm that MDB consumes few resources, that it is reliable, i.e., is able to reach most of the UEs, and that the latency involved is tolerable, even when the target area is quite large.

The rest of the paper is organized as follows: Section 2 reports background information. Section 3 reviews the related work. Section 4 discusses the role of UEs and eNBs in MDB. Section 5 reports performance evaluation results, and Section 6 concludes the paper.

2. Background

Hereafter, we describe the LTE-A protocol stack, as well as point-to-multipoint (P2MP) D2D communications.

An LTE-A network is composed of cells, under the control of a single eNB. UEs are attached to a eNB, and can change the serving eNB through a handover procedure. The eNBs can communicate between themselves using the X2 interface, a logical connection generally implemented on a wired network.

The LTE-A protocol stack incorporates a suite of four protocols, shown in Figure 1, which collectively make up layer 2 (i.e., the Data-link layer) of the OSI (Open System Interconnection) stack. The stack is present on both the eNB and the UE. Traversing the LTE-A stack from the top down, and assuming the viewpoint of the eNB, we first find the Packet Data Convergence Protocol (PDCP), which receives IP (Internet Protocol) datagrams, performs cyphering and numbering, and sends them to the Radio Link Control (RLC) layer. RLC Service Data Units (SDUs) are stored in the RLC buffer, and they are fetched by the underlying MAC (Media Access Control) layer when the latter needs to compose a subframe transmission. The RLC may be configured to work in three different modes: transparent (TM), unacknowledged (UM) or acknowledged (AM). The TM mode does not perform any operation. The UM, instead, segments and concatenates RLC SDUs to match the size requested by the MAC layer, on the transmission side. On the reception side, RLC-UM reassembles SDUs, it detects duplicates and performs reordering. The AM adds an ARQ (Automatic Repeat Request) retransmission mechanism on top of UM functionalities. The MAC assembles the RLC Protocol Data Units (PDUs) into Transmission Blocks (TB), adds a MAC header, and sends everything through the physical (PHY) layer for transmission.

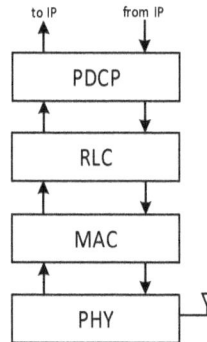

Figure 1. LTE-A (Long-Term Evolution-Advanced) protocol stack. IP: Internet Protocol; PDCP: Packet Data Convergence Protocol; RLC: Radio Link Control; MAC: Media Access Control; PHY: physical.

In LTE-A resources are scheduled by the eNB's MAC layer, at periods of a Transmission Time Interval (TTI) (1 ms). On each TTI, a vector of Resource Blocks (RBs) is allocated to backlogged UEs according to the desired scheduling policy. A TB may occupy a different number of RBs, depending on the Modulation and Coding Scheme (MCS) chosen by the eNB. The latter is selected based on the CQI reported by the UE, which is computed by the UE using proprietary algorithms, and corresponds to the Signal to Interference to Noise Ratio (SINR) perceived by the latter, over a scale of 0 (i.e., very poor) to 15 (i.e., optimal). Each CQI corresponds to a particular MCS, which in turn determines the number of bits that one RB can carry. Hereafter, we will often use the term CQI to refer to the (one and only) MCS that is determined by the former, trading a little accuracy for conciseness.

In the downlink (DL), the eNB transmits the TB to the destination UE on the allocated RBs. In the uplink (UL), the eNB issues transmission grants to UEs, specifying which RBs and which MCS each UE can use. In the UL, UEs need means to signal to the eNB that they have backlog. This is done both in band, by transmitting a Buffer Status Report (BSR) when scheduled, or out band, by starting a Random ACcess (RAC) procedure, to which the eNB reacts by issuing transmission grants in a future TTI. RAC requests from different UEs may collide at the eNB. To mitigate these collisions, the UEs select at random one in 64 preambles, and only RAC requests with the same preamble collide. After a RAC request, a UE sets a timer: If the timer expires without the eNB having sent a grant, that UE waits for a backoff period and re-iterates the requests.

The 3rd Generation Partnership Project (3GPP) has standardized Network-controlled D2D communications for LTE-A in release 12 [11]. These are point-to-multipoint (or one-to-many) communications having proximate UEs as the endpoints, i.e., without the need to use a two-hop path having the eNB as a relay. A D2D link is also called sidelink (SL). The SL is often allocated in the UL spectrum in a Frequency Division Duplex (FDD) system, since the latter can be expected to be less loaded than the DL one, due to the well-known traffic asymmetry [9]. Under this hypothesis, D2D-enabled UEs must be equipped with a Single-Carrier Frequency Division Multiple Access (SC-FDMA) receiver [18]. The phrase network-controlled hints at the fact that the eNB is still in control of resource allocation on the SL, i.e., it decides which UE can use which resources. Two schemes have been envisaged to do this: a Scheduled Resource Allocation (SRA), and an Autonomous Resource Selection (ARS). SRA is an on-demand scheme, similar to resource allocation in the UL for standard communications: the UE must send a RAC request to the eNB, which grants enough space for it to send its BSR. Then, the eNB schedules SL resources accordingly and issues the grant to the UE for D2D communications, as shown in Figure 2a. On the other hand, in ARS the eNB configures a static resource pool, e.g., M RBs every T TTIs, and UEs can draw from it without any signaling. With reference to Figure 2b, the UE has new data to transmit at $t = 1$, but it needs to wait for the next eligible TTI, i.e., at $t = 4$. If more than one UE selects the same resources, then collisions will ensue. Note that P2MP D2D transmissions are not acknowledged, hence the sender cannot know which neighboring UEs received a message, and H-ARQ (Hybrid Automatic Repeat reQuest) is disabled. This is because (N)ACKs should be sent on a dedicated control channel to the sender, but dimensioning the latter would be impossible: in fact, with P2MP D2D transmission, there is no way to know in advance how many and which UEs will actually receive the message, or were meant to in the first place. P2MP D2D transmissions use UM RLC.

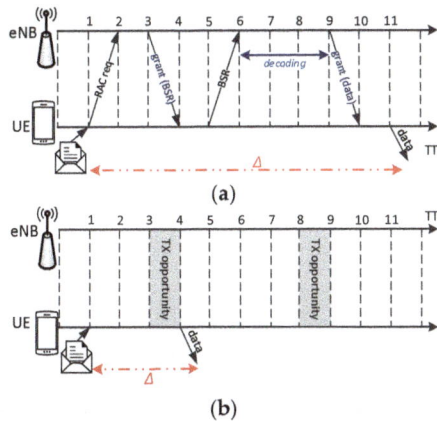

Figure 2. (**a**) Scheduled Resource Allocation (SRA); (**b**) Autonomous Resource Selection (ARS). Δ is the time between the generation of new data and the actual transmission. UE: User Equipment; eNB: eNodeB; RAC: Random ACcess; BSR: Buffer Status Report; TTI: Transmission Time Interval; TX: transmission.

3. Related Work

Multihop D2D communications in LTE-A networks have been studied in several works, e.g., [12–17]. In [12], authors consider multihop D2D communications in order to extend the network coverage and propose a resource allocation strategy to optimize the throughput along the multihop paths. The study is restricted to two-hop communications where one UE is identified as relay node for one pair of transmitting and receiving UE, where each hop is a unicast point-to-point (P2P) D2D transmission. The work in [13] proposes a theoretical formulation for computing the outage probability of multihop communications. Also in this case, P2P D2D transmissions are considered. An opportunistic multihop networking scheme for Machine-type Communications is presented in [14]. UEs exploits the Routing for Low-power and Lossy Networks (RPL) algorithm used in Wireless Sensor Networks (WSNs) to compute the best route toward a given destination. In [15], game theory is applied to find the best multihop path for uploading content from one UE to the eNB. Both [14,15] differ from our work as they deal with the problem of delivering messages toward a given destination, instead of disseminating them to all the UEs within a given target area. Moreover, P2MP communications are not considered. In [16], P2P D2D communications are considered in order to enhance the evolved Multimedia Broadcast and Multicast Services (eMBMS) provided by LTE-A networks. In this case, the eNB transmits its multimedia content to a subset of UEs and the latter exploit D2D links to forward the data to UEs with poor channel conditions in the downlink, i.e., cell-edge UEs. A similar problem is tackled in [17], where UEs receiving data from the eNB use one P2MP D2D transmissions to distribute the data to neighboring UEs. The above paper focuses on finding the best subset of relay UEs so that the total power consumption is minimized and the rate requirements for all the UEs are satisfied. None of the above works address the problem of disseminating UE-generated contents toward all UEs within a geographical neighborhood.

Centralized resource scheduling is sometimes assumed in Wireless Mesh Networks (WMNs), although far less often than distributed resource scheduling, (see, e.g., [19,20]). In this context, our broadcasting problem is superficially similar to the one of channel assignment and/or link scheduling in WMNs. However, the assumptions are quite different from those made in LTE-A, since nodes in a WMN are usually equipped with few radios, which can be tuned to a larger number of channels. In LTE-A, all UEs have as many "radios" and "channels" as the number of RBs, which is in the order of several tens. More importantly, RBs can be allocated dynamically to UEs, whereas the algorithms presented in the literature often assume periodic transmissions and long-term, semi-static resource allocations. Moreover, unicast P2P transmissions are considered. For these reasons, the broadcasting problem considered in this paper cannot be accommodated using the above algorithms.

Broadcast diffusion problems have been addressed in the context of mobile ad-hoc networks (MANETs) (e.g., [21,22]), especially to support the dissemination of routing alerts or for gossiping applications [23]. Unlike LTE-A, where resources are centrally scheduled by the eNB on demand, the above networks are infrastructureless and have distributed resource allocation. Work [24] reviews and classifies the broadcasting methods in MANETs, focusing on how they try to limit collisions, the latter being the key issue in an infrastructureless network. For example, [25] employs similar hypotheses as this work (no knowledge of the underlying topology, fixed transmission range), although in a different technology, and proposes a method for limiting the number of broadcast relaying, and thus of collisions, by preventing nodes transmission based on an adaptive function of the number of received copies of the same message and the number of its neighbors. The considered function can then be tuned to trade user reachability with broadcast latency. We show in Section 5 that the latency of MDB is in line with the ones of [25], but the delivery ratios are higher, in similar scenarios. One of the main purposes of this paper is in fact to show that MDB can leverage LTE's centralized scheduling. The combination of centralized scheduling and distributed transmissions is in fact unique to D2D-enabled LTE-A. Note that the ASR mode, described in Section 2, does instead allow unscheduled, collision-prone medium access, similar to what a MANET would do. In Section 5, we will show that such collisions actually hamper the performance.

4. Multihop D2D Broadcasting

In the following, we consider an LTE-A system composed of several cells, where UEs are D2D-enabled. UEs run applications that may generate messages (e.g., vehicular collision alerts) destined to all other UEs running the same application, within an arbitrary target area. Our problem is to reach as many interested UEs in the target area as possible, using only P2MP D2D transmissions, relayed by UEs themselves, using as few resources as possible. The system model is shown in Figure 3, where the shaded UE originates a message that has to be delivered to all the UEs within the circle. The solid arrows represent the first P2MP D2D transmission, and the dashed ones represent transmissions relayed by UEs in the first-hop neighborhood of the originating UE. A UE that perceives collisions in the same time/frequency resources will still attempt to decode the message received with the strongest power, i.e., it will exploit the so-called capture effect, typical of wireless networks [26].

In multihop D2D broadcast, the eNB does not participate in data plane transmissions, i.e., it does not send data packets. Data-plane transmissions are instead performed by the UEs themselves, on behalf of the applications running on them. However, the eNB still controls the resource allocation, hence can affect the performance of the broadcast. We only assume that the eNB allocates resources for generic D2D transmissions using standard-compliant means (to be discussed later in this section), unaware of the fact that multihop relaying is going on for D2D transmissions, or of specific application requirements (e.g., deadlines, target areas, etc.). In other words, we assume minimal, standard-compliant support from the infrastructure.

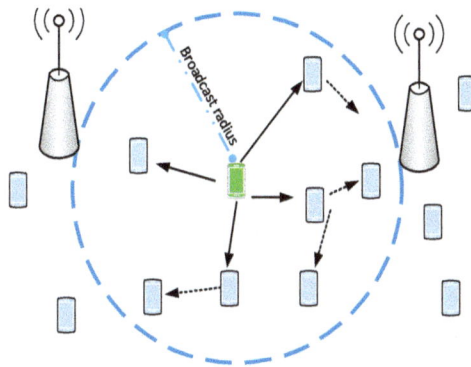

Figure 3. System model.

Multihop D2D broadcasting requires that applications decide which UEs to target and how to relay messages, whereas the LTE-A network allocates resources for D2D transmissions to allow multihop relaying. This implicitly assumes a trusted environment, where UEs behave cooperatively. Security is notoriously difficult to enforce in broadcast networks, and we refer the interested reader to [27] for a discussion of security issues in D2D communications in particular. Regarding cooperation, UEs may be inclined to behave selfishly to save battery or to avoid increasing traffic volumes in pay-per-use plans. On one hand, suitable incentives and reputation-based schemes, such as those discussed in [28,29], could mitigate the problem. On the other hand, MDB services are supposed to be used also by embedded applications (e.g., application software running on cars), which have access to an energy source (e.g., the vehicle's battery) and are not under the control of the end user (i.e., the car owner or pilot). In this last case, manufacturers would clearly benefit from coding cooperative behaviors in their embedded software. Hereafter, we assume that the UEs running MDB behave cooperatively. In Section 5 we evaluate the impact of selfish users on the performance.

Hereafter, we first discuss how UE applications should be designed in order to support broadcasting effectively, and then move on to discussing resource allocation policies in the network.

4.1. Broadcast Management within the UE

The two problems that UE applications should solve are: (i) how to identify the set of potential recipients; and (ii) when to relay D2D communications. The first problem boils down to identifying all the UEs running the same application in a certain geographical area. UEs running the same application can register to a reserved multicast IP address. This is relatively easy to do with IPv6 (Internet Protocol Version 6), where multicast address format is flexible. As far as defining the target area is concerned, we argue that the area depends on both the network scenario and the application: in a vehicular use case, for instance, vehicle collision alerts should reach vehicles in a radius of few hundred meters, whereas traffic notifications should probably travel larger distances, allowing drivers to route around congested areas. This means that the application message should contain enough information to allow a recipient UE to understand whether or not it should relay it. The information regarding the target area should then be embedded in the application-level message. A simple, but coarse, approach to do this is via a Time-to-live (TTL) field: the source UE sets the TTL in the application message to a desired maximum number of hops. Each relaying UE, then, decreases that field by one, and relays the message only if TTL > 0. While this is relatively simple and economical in space (an 8-bit field should be enough for must purposes), the downside is that the source UE can exert little control over the area covered by the broadcast, since the latter ends up depending on both radio parameters (such as the UEs' CQI and their transmission power) and network topology (i.e., the position and density of UEs). The latter, in turn, is unpredictable and changes over time, so that any default value runs the risk of being too small or too high. The alternative is to code the target area within the message, by inserting the originating UE's coordinates and the boundaries of the target area. Geographic coordinates can be taken from GPS positioning, or from geolocation services co-located with the network (e.g., using MEC solutions [8]). Geographic coordinates can be represented by two 32-bit floating points, indicating latitude and longitude with enough precision. A simple way to constrain the target area is to encode a maximum target radius, thus making it circular. Assuming that the target radius is represented in meters, a 16-bit integer should be large enough for most purposes. Encoding originator's coordinates and target radius allows one to define with more precision the target area, which becomes independent of the UE's density and location. This comes at the cost of using more space in the message (10 bytes overall instead of one). Increasing the message size, in turn, entails consuming more network resources for transmission. Obviously, more advanced definitions of target areas can also be envisaged, at the cost of further increasing the message payload. With a geographical representation of the broadcast domain, receiving UEs can then check whether their own position falls within the target area before relaying the message. This can be done by using simple floating-point arithmetic, i.e., by computing the distance from the originator and checking whether it is smaller than the target radius included in the message. Given the coordinates of the two points A and B (specified in latitude φ and longitude λ), the Haversine formula [30] is used to compute the shortest distance over the earth's surface, which is:

$$d = 2r \sin^{-1}\left(\sqrt{\sin^2\left(\frac{\varphi_B - \varphi_A}{2}\right) + \cos(\varphi_A)\cos(\varphi_B)\sin^2\left(\frac{\lambda_B - \lambda_A}{2}\right)}\right) \tag{1}$$

where r is the earth's radius.

Note that using a geographical representation (even one with infinite precision) still leaves a margins of uncertainty as to which UEs will receive the message: in fact any UE which is inside the target radius will relay the message, hence all UEs within an annulus of one D2D transmission radius outside the edge of the target area may still receive it.

UE applications should also take care of relaying. In fact, it is at the application level that suitable algorithms can be run to make relaying efficient. A brute-force relaying, whereby UEs relay all received messages, would in fact quickly congest the network, since the same UE would receive the message from several neighbors, and relay them all unnecessarily. This would waste resources that could otherwise be used for other purposes. In order to make relaying efficient, a suppression mechanism can be used, e.g., the one of the Trickle algorithm [31]. Trickle is used in WSNs to regulate the relaying of updates and/or routing information. In that context, Trickle runs on each node participating in the broadcasting: before sending a message, the node listens to the shared medium in order to figure out if that information is redundant, i.e., enough neighboring nodes are already sharing it. If so, it abstains from transmission so as to avoid network flooding. In Trickle, two parameters can be configured: the Trickle Interval I and a number of duplicates K_{max}. A UE selects a random time window $\Delta_{trickle}$ in $[I/2, I)$, and counts the copies of the same message received therein. The UE only relays a message if it receives fewer than K_{max} copies of it. To sum up, Figure 4 depicts the flowchart of the operations performed by a UE application on reception of a message, when the Trickle suppression algorithm is employed. First, the UE checks whether the incoming message had already been received. If so, it abstains from relaying and the procedure terminates. Otherwise, it computes the distance from the originating UE and compares it with the maximum target radius. If the UE is inside the target area, then Trickle operations are initiated: the Trickle timer is started and the duplicate counter k is set to 0. Figure 5 shows that k increases on each duplicate reception within $\Delta_{trickle}$. When the above timer expires, the UE relays the message if $k < K_{max}$.

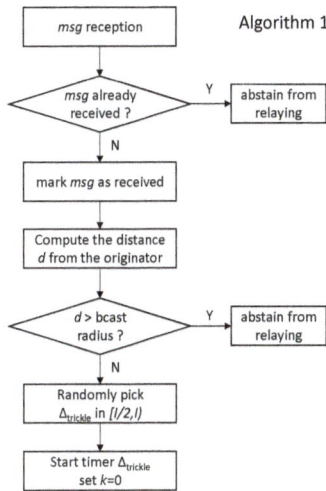

Figure 4. Flowchart of UE-side operations at message (msg) reception.

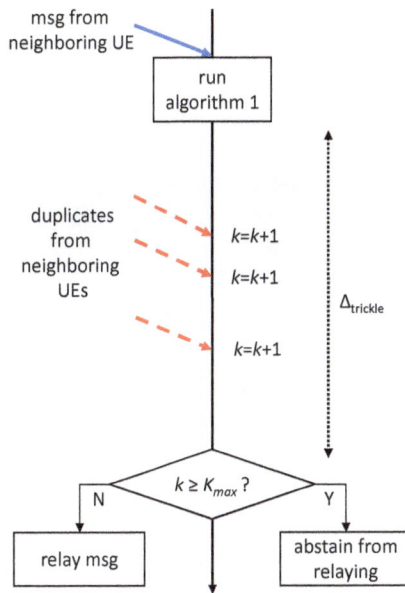

Figure 5. Relaying operations at the UE application.

Note that, in the absence of Trickle, UEs must be made to wait for a random time chosen uniformly in $[0, \Delta)$ before attempting a relay. Parameter Δ can be tuned to trade collision probability for latency. In fact, P2MP D2D transmissions will reach several UEs simultaneously. In the absence of random delays, these UEs will attempt relaying at the same time, since LTE-A is slotted. This may lead to collisions, regardless of how the eNB allocates resources (an issue which is dealt with in the next subsection).

So far we have assumed that a message is originated by one UE. However, messages are supposed to be generated as reactions to events (e.g., a vehicular collision), and the same event may be detected by multiple UEs, which might then initiate a broadcast quasi-simultaneously. In fact, whichever the allocation scheme in the network, there is a time window where all UEs that want to start a broadcast will be unaware of others doing so, even if they are within D2D hearing range of one another. That time window is around 10 ms with SRA (i.e., the time it takes to complete a resource allocation handshake), and equal to the period with ARS. What happens in this case depends on how the application handles the different broadcasts. A baseline solution is to do nothing. In that case, since the information included in the messages is not the same, since e.g., originators' coordinates are different, the latter will be considered as different broadcasts by the Trickle instances running in the UEs' applications, and will be broadcast independently. As a result, multiple, independent broadcasts related to the same event will traverse the network, with a corresponding increase of the traffic load. On the other hand, the applications running at the UEs can easily be endowed with the necessary intelligence to associate two (or more) messages, possibly with a different payload, to the same event: for example, if the distance between the originators' is below a threshold, the message type is the same, and the reception times are again within a predefined window (which may be computed based on the Trickle window). In this case, merging can occur, i.e., the various broadcasts messages are associated to the same Trickle instance, thus being perceived as duplicates of the same broadcast process. Note that MDB can also accommodate messages generated by entities other than UEs, e.g., nodes located in the Internet, the LTE core network, or—possibly—Mobile-edge Computing servers running applications on behalf of the UEs. In that case, the network can select one (or more) proxy originator UE(s) and send them the message from the serving eNB(s) using downlink transmissions. The receiving UE(s) can, in turn, initiate the broadcasting procedure as described above.

4.2. Resource Allocation in the Network

As discussed in Section 2, the eNB controls resource allocation, and may use either SRA or ARS. We now compare the two approaches, highlighting their pros and cons in the context of multihop relaying, also taking into account that multicell relaying may be required.

As far as latency is concerned, using SRA requires each UE to undergo one RAC handshake per transmission. As shown in Figure 2a, this handshake takes a 10 ms delay in the best of cases, i.e., when the eNB issues grants immediately. If RAC collisions are experienced, or the eNB delays scheduling because the UL is congested, the per-hop delay may be even larger. On the other hand, with ASR, a UE can send a message as soon as a transmission opportunity becomes available, without the need of going through a RAC/BSR handshake. Thus, with ARS the maximum scheduling delay is given by the resource allocation period T. Using ASR, especially with small periods, allows faster access to the medium. However, this entails allocating a large share of resources to P2MP D2D transmissions statically, thus wasting resources when these are not required, and preventing standard UL communications to use them. Therefore, with ASR, latency is traded off for resource efficiency.

The two allocation schemes differ greatly regarding collisions. When using SRA, the only possible collisions are those of simultaneous RAC requests at the eNB. However, these are quite unlikely. The LTE-A standard requires UEs to select a preamble among 64 possible choices. Simultaneous RAC requests with different preambles do not collide. Furthermore, when a RAC request is not answered by the eNB (either because of a RAC collision or because the eNB does not have resources to spare), the UE simply sends it again after a backoff time. Thus, RAC collisions do delay the broadcast process, but they also desynchronize relaying UEs, which is a positive side effect. With SRA, data transmission on the SL is instead interference-free, since the eNB generally grants SL resources to one transmitting UE at a time. The only exception to that rule is when the eNB exploits a frequency reuse scheme (such as the one in [32]), in which case faraway, non-interfering UEs may be granted the same RBs simultaneously. However, this happens exactly because the eNB knows that they will not interfere with each other. If, instead, ARS allocation is used, UEs claim RBs on the SL for their own transmission without a central scheduling and without their neighbors knowing, hence the intended receivers face unpredictable interference. The latter can be mitigated by having the UEs select at random which

RBs to use, and by dedicating more resources to SL transmissions, which decreases the efficiency. Moreover, ARS allocation is periodic, hence it implicitly forces synchronization among groups of UEs: all UEs whose application requests a relay in the same ARS period will end up accessing the SL at the next ARS opportunity, hence increasing the likelihood of collisions. This would happen at each hop. Furthermore, since a sender does not know if collisions have occurred, the only possible countermeasure to increase the reliability of a transmission would be to retransmit the same message more than once.

As already discussed, a target area may include more than one cell, as shown in Figure 6. This poses the problem of coordinated resource allocation among neighboring cells. In fact, if each cell allocates SL resources autonomously, cell-border UEs will be subject to interference from UL transmissions in the neighboring cells, hence they may be unable to receive P2MP D2D transmissions. This problem is likely to affect more heavily dense networks [33], where cells are smaller.

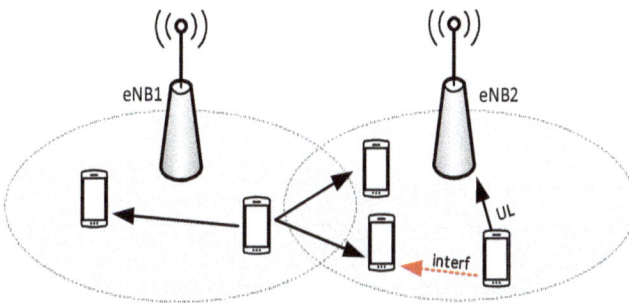

Figure 6. Multicell scenario. UL: uplink.

If the network uses ARS allocation, coordination is fairly easy to achieve: all it takes is that neighboring cells use the same allocation pattern. If, instead, SRA is used, resources are allocated on demand, hence the eNB must share information regarding their allocation using the X2 interface. For instance, an eNB may inform its neighbor(s) about which RBs will be allocated to a cell-border P2MP D2D transmission in a future TTI, so that the neighboring eNB(s) avoid allocating the same resources to UL or D2D transmissions in the vicinity of the cell border. This requires the sending eNB to plan scheduling on the SL (at least for cell-border UEs) with a lookahead of some TTIs, enough for the above message to reach its neighbor through the X2. With reference to Figure 7, at $t = 9$ the eNB informs its neighbor that, in a future TTI, a grant for a cell-border P2MP D2D transmission will be scheduled on a given set of RBs. The receiving eNB marks the advertised RBs as occupied at the appointed TTI, and performs its usual scheduling. The lookahead mechanism can be expected to add a negligible delay to the broadcast diffusion, since the X2 connection is normally wired and low-delay.

The eNBs should also select the MCS of P2MP transmissions. Such choice should strike a tradeoff between two conflicting objectives, i.e., transmission range and resource consumption. In fact, selecting more performing MCSs implies reducing the number of RBs required for a transmission, since more bits will be packed in the same space. However, it will also decrease the transmission range, since the distance at which the SINR will be high enough to allow successful decoding decreases with the CQI. This implies that more hops will be required to cover a given target area. Conversely, selecting less performing MCSs will require fewer hops, but more RBs per transmission. Note that the eNB must choose the MCS only if it uses SRA allocation: in this case, in fact, the eNB sends D2D grants, which carry indication of the transmission format. If, instead, ARS is used, UEs may select the MCS autonomously, at least in principle. In practice, we argue that the eNB should still make that choice, and possibly advertise it periodically using RRC procedures. In fact, the eNB is in a better position than single UEs to assess the UE density or location, hence to select the most suitable cell-wide transmission format.

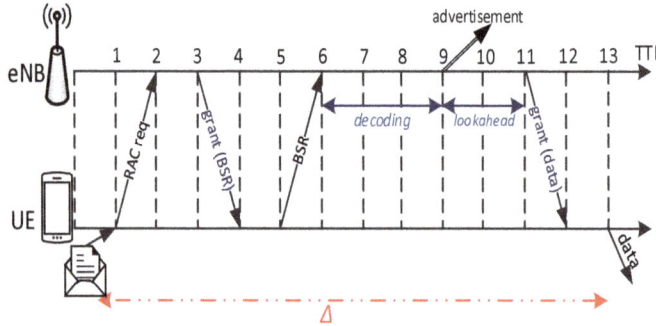

Figure 7. SRA with coordination among eNBs.

5. Performance Evaluation

In this section, we employ system-level simulations to assess the performance of MDB services. To do so, we use SimuLTE [34], a simulation framework based on OMNeT++ [35], which models both eNBs and UEs endowed with a Network Interface Card (NIC) that implements the data plane of the LTE-A protocol stack, from application- to physical-layer. In particular, it allows one to simulate both P2P and P2MP D2D communications [36]. For the scope of this paper, we enhanced the simulator with a new application module running at the UE side, which is able to send messages to neighboring UEs, leveraging P2MP D2D communications provided by the underlying LTE-A NIC. The receiving UE can in turn relay the message using again P2MP D2D communications. In the following, we consider a first scenario where UEs are static and we evaluate the impact of the different application- and MAC-level mechanisms described in Section 4. Then, we will investigate a vehicular network scenario, where UEs are mobile, as a use-case for MDB.

5.1. Tuning of the Multihop P2MP Settings

Our first simulation scenario, shown in Figure 8, includes five adjacent eNBs, located at a distance of 400 m from each other. We initially assume that each eNB serves 30 UEs with P2MP D2D capabilities, which are deployed randomly in a narrow strip along a straight line. UEs are assumed to be static. Their transmission power is 30 dBm in the UL and 15 dBm in the SL. The channel model includes Jakes fading and log-normal shadowing. Table 1 reports physical-layer parameters. We assume that broadcast messages transport a 4 byte payload, representing a code for indicating the type of the message. Considering the additional information for the originator's coordinates and the target radius discussed in Section 4, we assume a total length of 14 bytes. A message is generated by a random UE every second, starting a new broadcast. As we will show later, the duration of each broadcast is less than one second, hence we can consider each message dissemination as an independent event. For each configuration, we run one instance of 100 s and statistics are obtained by averaging 100 independent broadcasts. Confidence intervals at the 95% level are shown. In the following, we assume that UEs relay the same message only once and that the target radius is 1000 m, unless otherwise specified.

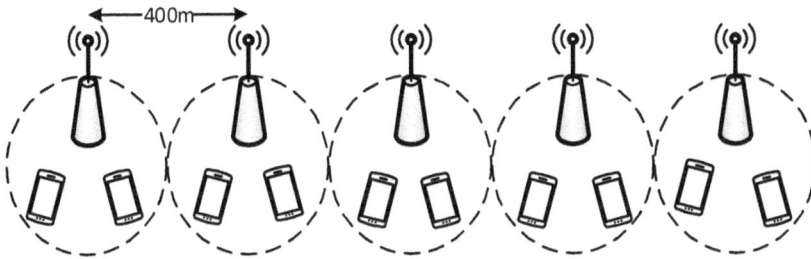

Figure 8. Evaluation scenario.

Table 1. Physical-layer parameters. ITU: International Telecommunication Union; eNB: eNodeB; UE: User Equipment; UL: uplink; D2D: device-to-device.

Parameter	Value
Carrier frequency	2 GHz
Bandwidth	10 MHz (50 RBs)
Fading model	Jakes
Path loss model	ITU Urban Macro [37]
Noise figure	5 dB
Cable loss	2 dB
eNB Transmission Power	46 dBm
UE Transmission Power-UL	30 dBm
UE Transmission Power-D2D	15 dBm

5.1.1. Varying Application-Level Settings

In this subsection, we evaluate how different settings of the UE's application layer affect the performance of the broadcasting and compare MDB with the relaying made by the eNB. We assume that the eNB uses the SRA allocation policy, using the scheduling algorithm described in [32], where two or more UEs can reuse the same RBs if they are not interfering each other. In particular, interference conditions are modeled through a conflict graph [38], which is maintained by the eNBs according to UEs' location information. In the conflict graph, two UEs are conflicting if the power they receive from each other is above a threshold of −50 dBm. In that case, the UEs are placed on different RBs, otherwise they can share the same RBs.

Figure 9 shows which UEs receive a broadcast message, when either TTL or GPS coordinates and radius are used. The UE marked with the cross is the one that originates the message, and the boundary of the target area (i.e., the shaded one) is marked by the vertical dashed line. Note that the line only looks straight because the scale on the y-axis is stretched. UEs that received the message are shown as green circles, whereas red triangles represent UEs that did not. Using coordinates and radius (top left of Figure 9) allows the message to reach all the desired UEs, and few of those outside the range. When TTL is used, the set of UEs depends on the initial TTL value. If the value is set to 5 or 6, some UEs at the border do not receive the message, whereas setting it to 7 covers the entire target area. However, if the line is moved in either direction, or the UE density or the network MCS change, that value stops being optimal and must be recomputed. In order to exemplify the dissemination process of the message, Figure 10 shows the per-TTI allocation of RBs at the eNBs involved in the broadcasting process, which is started by a UE served by eNB1. Thus, the latter allocates RBs to UEs under its control to let them relay the message. After some time, the message reaches UEs served by eNB2 and eNB3, which in turn start granting transmission resources to them. eNB4 and eNB5 do not allocate RBs, since their served UEs are outside the target area.

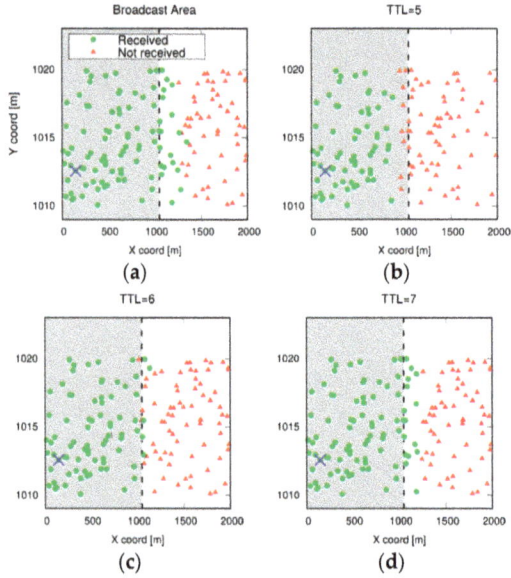

Figure 9. Target area vs. TTL.

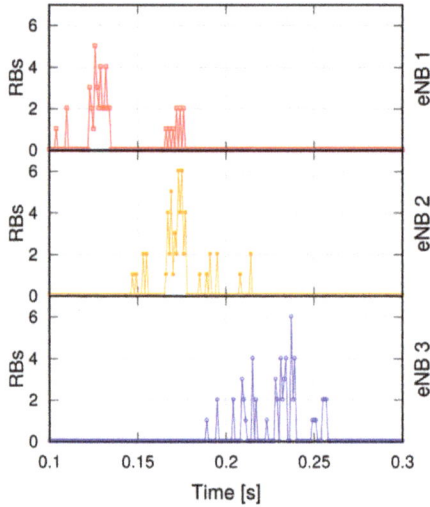

Figure 10. RB (Resource Block) allocation at different eNBs over time.

We now evaluate the impact of the Trickle suppression mechanism on MDB. In particular, we consider different settings for its relevant parameters K and I. We recall that a UE relays a message if fewer than K duplicates are received within a time randomly selected in the range $[I/2, I)$, I being the Trickle interval. Defining S as the number of transmissions performed by all UEs in the floorplan to relay a single broadcast, Figure 11 shows $E[S]$, whereas Figure 12 reports the 95th percentile of the time required to complete the dissemination, computed as $t_e - t_0$, where t_0 is the time at which the originating UEs starts the broadcast, and t_e is the time at which none of the UEs that have received the message can forward it anymore (either because they already have, or because Trickle suppressed relaying).

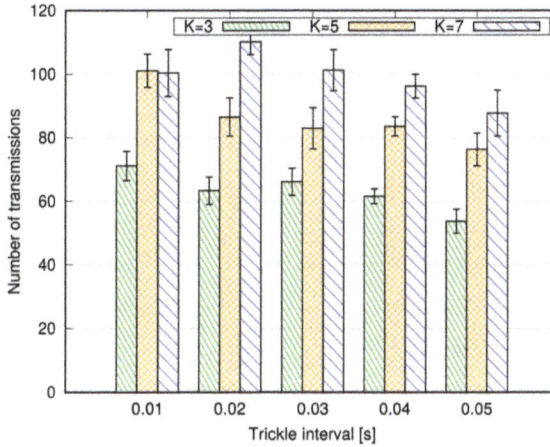

Figure 11. Average number of application-level transmissions with different settings of the Trickle algorithm.

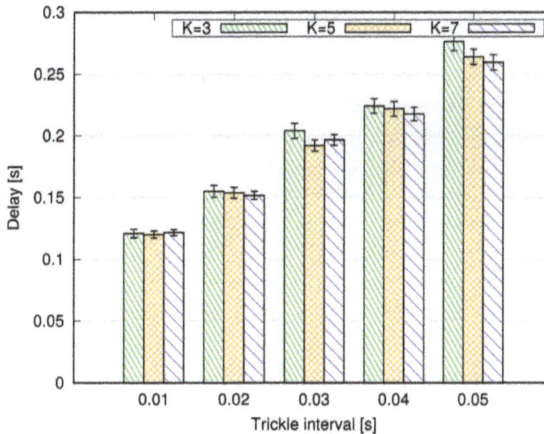

Figure 12. 95th percentile of application-level delay with different settings of the Trickle algorithm.

We observe that the combination of large values of I and small values of K allows us to transmit fewer messages, hence saving in terms of radio resources and power at the UEs. Of course, larger Trickle intervals result in larger delays, since UEs wait longer before relaying a message. The value of K has a negligible impact on the latency of the broadcasting. If our primary objective is to provide fast diffusion of a message, we should then select a short Trickle interval and a small value of the threshold K.

For the above reasons, we now set the Trickle parameters to $K = 3$ and $I = 10$ ms, and compare the performance of the with and without Trickle broadcasts. Figure 13, left, shows that adding Trickle increases the mean delay: in fact, Trickle adds delay before a relaying is attempted, and prevents some UEs from relaying the message at all. These delays add up at each hop. However, Trickle allows significant resource savings, quantifiable in about half the RBs.

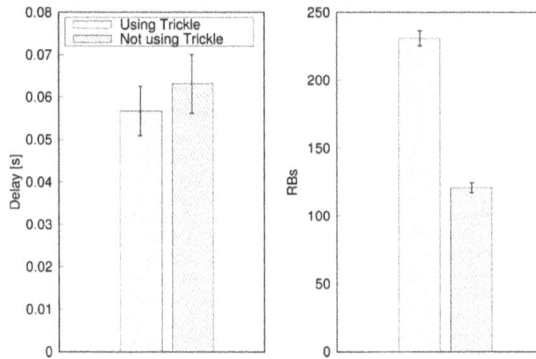

Figure 13. Average delay (**left**) and average allocated RBs (**right**), w and w/o Trickle.

We then evaluate the performance of MDB against unicast eNB relaying. As far as the latter is concerned, we assume that the source UE sends the message to the eNB, which relays it within its cell using unicast DL transmissions, and sends it to its neighboring eNB using the X2 interface. We envisage two options to geofence the broadcast: one (which is called eNB relaying) assumes that the eNBs relay the message to all UEs in their cell, hence the target area consists of a number of cells. The other solution (called enhanced eNB relaying) consists in selecting the subset of UEs to target in each cell based on their geographic position. In both cases, the eNBs must be endowed with an application layer that sends and receives messages. With eNB relaying, the eNB application in the originating cell must read the message payload, and—specifically—the GPS coordinates of the source UE and the target radius, to understand which other neighbors to contact, if any. The eNB applications in the other cells will just receive the message and request the unicast forwarding. With the enhanced eNB relaying, instead, all eNB applications must also read the GPS coordinates and radius, find out which UEs are within the target area, and request the unicast relaying. This entails knowing the position of the UEs in the cell. As already stated, this can be achieved leveraging MEC solutions, but it can be expected to have a cost, in terms of added communication latency and overhead.

Figure 14 reports the 95th percentile of the delay and the allocated RBs in the DL subframe. As expected, the broadcast is completed sooner using unicast relaying, since UEs can be reached in two radio hops plus, possibly, a (fast) X2 traversal. However, the cost in terms of allocated resources is non-negligible: a broadcasting occupies 230 RBs in the DL subframe, whereas MDB does not require the eNB to use the DL spectrum at all. Note that, in a 10 MHz deployment, using 50 RBs per TTI, unicast relaying would stall the DL for 4–5 consecutive TTIs, which is unadvisable.

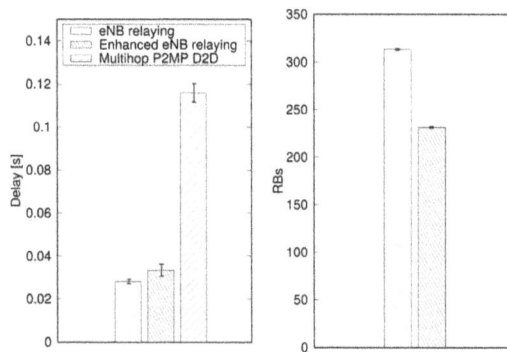

Figure 14. 95th percentile of delay (**left**) and average allocated RBs in downlink (DL) (**right**).

5.1.2. Varying MAC-Level Settings at the eNB

In the following, we assume that the target area is embedded in the message, and Trickle suppression is enabled.

We first discuss the tradeoff involved in the choice of the MCS at the cell level. In Figure 15, the x axis reports the mean of the total number of RBs per broadcast, whereas the y axis shows the mean reception delay of UEs within the target radius. Resources are allocated via SRA, and the points represent the CQIs (which determine the MCS). As expected, the number of allocated RBs decrease with the CQI. In fact, higher CQIs correspond to more performing MCSs, hence to more bits per RB. A message transmission occupies 11 RBs with CQI 3, and one RB with CQI 15. However, using larger CQIs increases the number of hops, hence the latency. This is evident in the two segments 7–9 and 10–15, where the same number of RBs is used, but larger latencies are obtained, since the reception range is reduced. However, latency also increases with too small CQIs. As previously discussed in Section 3, the probability of correct reception decreases with the number of RBs employed, due to frequency-selective fading, all else being equal. The MCS corresponding to CQI 7 strikes a good tradeoff between latency and resource occupancy. While the absolute values in the graph are a function of the target radius, qualitatively similar results (although on different scales on both axes) are obtained when the radius is varied. From now on, CQI 7 is used in the simulations.

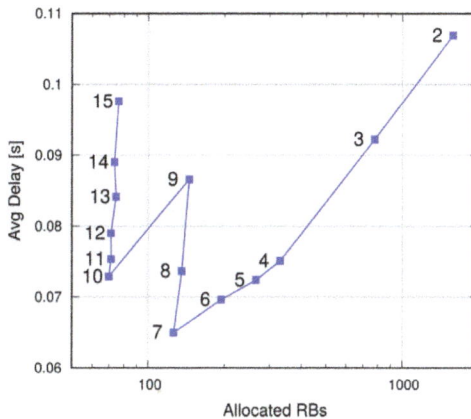

Figure 15. Average allocated RBs and average delay as a function of CQI (Channel Quality Indicator).

We now evaluate the two resource allocation strategies. Figures 16 and 17 report the average and the 95th percentile of the delay, for different target radiuses, achieved with SRA and ARS. We configured ARS with four different patterns, consisting of 20 RBs allocated at periods of 5, 10, 20 and 50 ms respectively. When the period is small enough (i.e., 10 ms or below), ARS is faster than SRA, since the time to the next transmission opportunity is smaller than the duration of a RAC handshake. This comes at the price of higher resource consumption: ARS consumes many more resources, which go unused when there is no traffic. For instance, with a period of 5 ms, one eNB must reserve 8% of its resources for P2MP D2D broadcasts (still assuming a 10 MHz deployment). SRA, instead, only uses the RBs requested by the UEs, which are around 130 per broadcast over a 1000-m radius (involving 5 eNBs), also factoring in BSR transmissions, as shown in Figure 18. A back-of-the-envelope computation shows that, since UEs send one broadcast per second, then the resource occupancy of SRA is less than 130 RBs per second in the whole network, against 10 thousand for ARS at 5 ms.

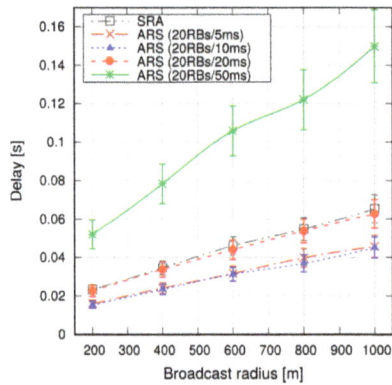

Figure 16. SRA vs. ARS, average delay.

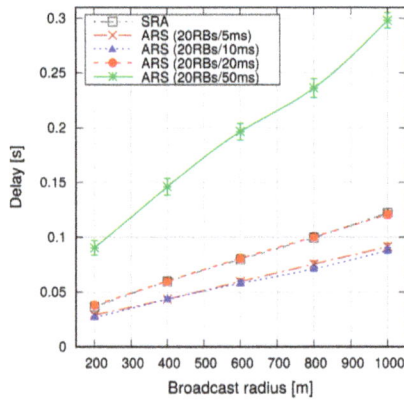

Figure 17. SRA vs. ARS, 95th percentile of delay.

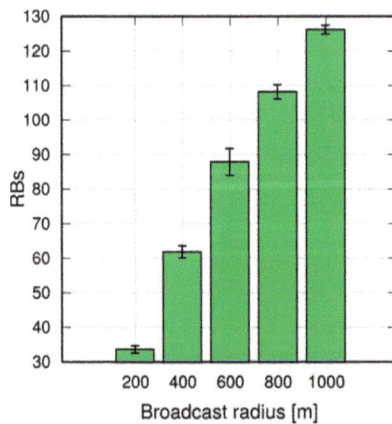

Figure 18. SRA, avg. allocated RBs per broadcast.

Figure 19 reports the percentage of UEs that actually receive the message, which is close to 100% and fairly insensitive to the target radius. Note that the last result is partly due to the spatial reuse policy, which allows an eNB to schedule more than one D2D transmission on the same time/frequency resources.

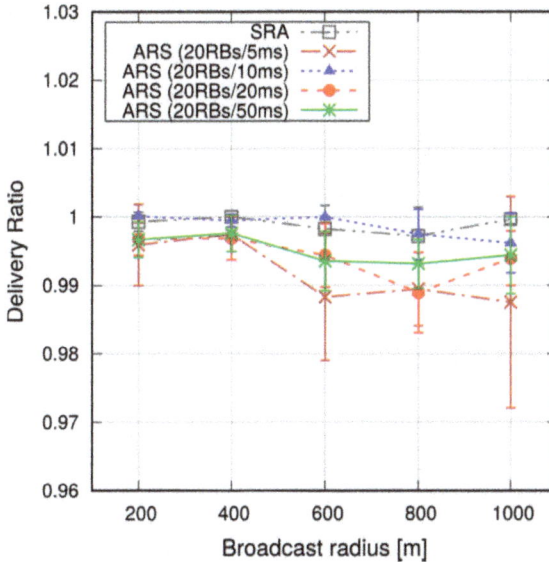

Figure 19. SRA vs. ARS, delivery ratio.

To put the above results into context, we observe that the average broadcast latency with SRA and a radius of 800 m is around 50 ms (see Figure 16), and the corresponding delivery ratio is above 99% (Figure 19). Compared to [25], which addresses broadcasts in an 802.11-based MANET in a somewhat similar scenario (although similarities between such different technologies as 802.11 and LTE are to be taken with a pinch of salt), we note that the latencies are similar, but our delivery ratio is remarkably higher (99% against 90–98%), despite the fact that LTE relies on much smaller UE D2D transmission radiuses (100–150 m against 500 m in [25]).

From now on, we use SRA as the resource allocation scheme, since it achieves a good tradeoff between latency and resource consumption.

5.1.3. Varying the Network Scenario

We now show what happens when network conditions are modified. More specifically, we show what happens if quasi-simultaneous broadcasts are started at nearby UEs (which may register the same event, e.g., a collision, and start dissemination independently). Moreover, we discuss how the performance of MDB is affected by UE density in the network, and the impact of selfish users.

In order to assess the impact of multiple originators for an event, we perform simulations where two broadcasts are started simultaneously by two UEs located at a maximum distance of 20 m. Figure 20 shows that the 95th percentile of the delay is slightly smaller than in the case where the broadcast has a single originator. This is because two UEs transmit simultaneously at the first hop and their messages can be possibly received by a few more UEs than in the single-originator case. On the other hand, merging makes little difference as far as delay is concerned, once we assume two originators. The effects of merging two broadcasts into a single Trickle instance are instead visible in terms of allocated RBs, as shown in Figure 21. As expected, having two independent broadcasts doubles the number of allocated resources, whereas the RB occupation in the case of merged broadcasts is essentially the same as the case with a single originator.

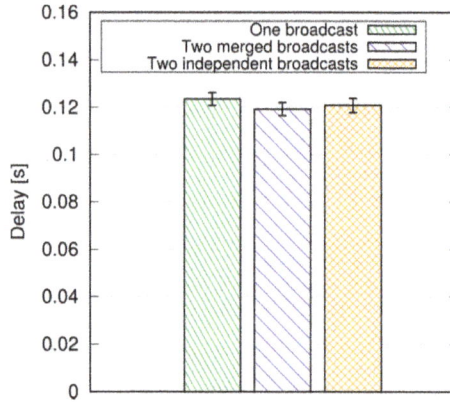

Figure 20. 95th percentile of delay with multiple broadcast sources.

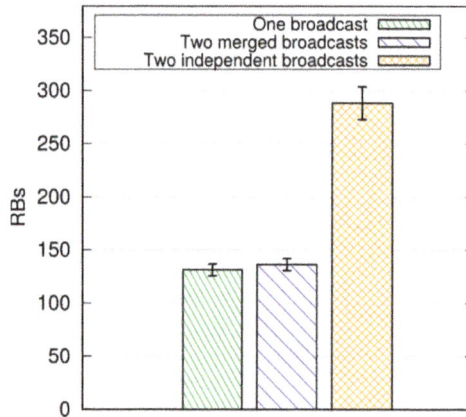

Figure 21. Avg. allocated RBs per broadcast with multiple broadcast sources.

We now assess the performance of MDB in denser networks. We increase the number of UEs served by each eNB from 30 to 150, resulting in a total number of UEs of up to 750. Again, we use a target area with a 1000 m radius. Figure 22 shows the 95th percentile of the delay. We observe that the delay decreases with the UE density, and it is slightly above 80 ms with 750 UEs. This is explained by the larger number of UEs receiving a single broadcast message, hence a larger number of potential relays. This in turn increases the probability of reaching farther UEs. Clearly, more transmissions come at the cost of a higher resource consumption, as reported in Figure 23: the number of allocated RBs per broadcast increase from 130 RBs (with 150 UEs) to about 400 RBs with 750 UEs. Interestingly, the ratio between the allocated RBs and the number of UEs decreases. In fact, more transmissions means more duplicates too, hence more UEs abstain from transmission thanks to the Trickle suppression mechanism.

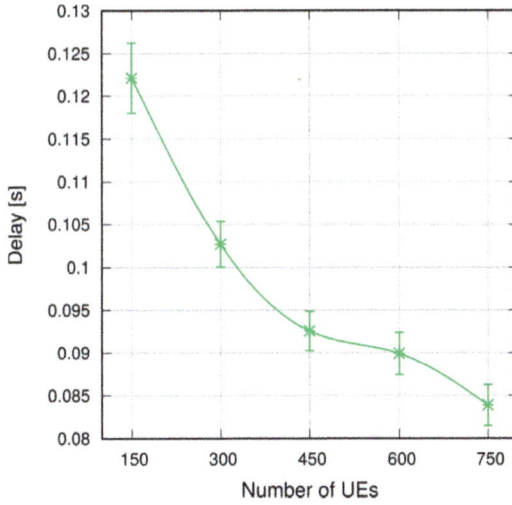

Figure 22. 95th percentile of delay, dense scenarios.

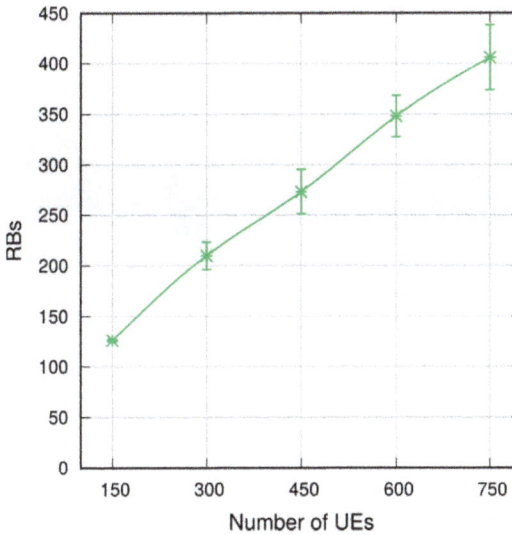

Figure 23. Avg. allocated RBs per broadcast, dense scenarios.

As discussed in Section 4, some selfish UEs may refuse to relay messages, e.g., to save their battery. In order to assess the impact of selfish users in MDB, we simulated scenarios with an increasing percentage of selfish UEs, at different densities. Figures 24 and 25 show respectively the 95th percentile of the delay and the number of RBs allocated per broadcast. As expected, increasing the percentage of selfish UEs results in a larger, though tolerable, delay. This is due to the reduced number of transmissions, which in turn also reduce the number of allocated RBs. In any case, the reliable delivery of the message is ensured by the Trickle suppression mechanism: in fact, (cooperative) UEs relay the messages if they perceive that their doing so is necessary to its diffusion. It is worth noting that the results obtained with 300 UEs and 50% of non-cooperative UEs (rightmost end of the dotted line in Figures 24 and 25) are similar to those obtained with 150 UEs and no selfish UEs (leftmost end of the continuous line in Figures 24 and 25). All things considered, the presence of randomly placed selfish users can be accounted for as an equivalent reduction in user density as far as latency and RB utilization are concerned.

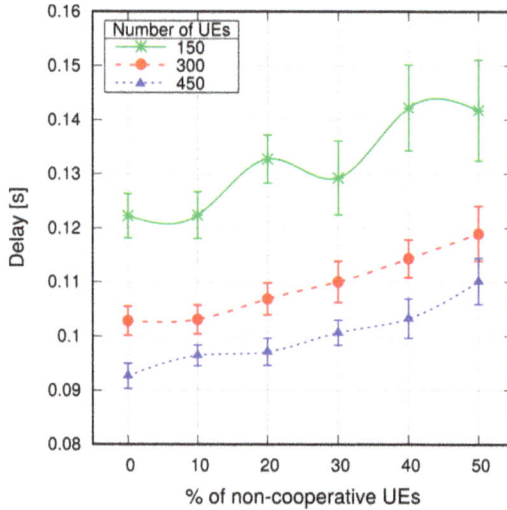

Figure 24. 95th percentile of delay with non-cooperative UEs.

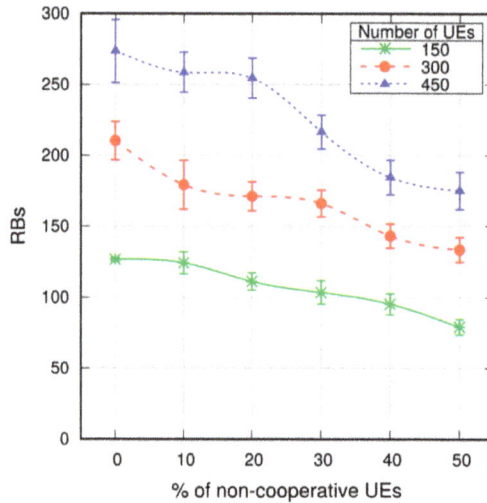

Figure 25. Avg. allocated RBs per broadcast with non-cooperative UEs.

5.2. Multihop D2D Broadcasting in Urban Vehicular Networks

We now discuss how MDB performs in a vehicular network. In this scenario, moving vehicles should be able to exchange information (like alert messages or traffic updates) with other vehicles, roadside elements (e.g., traffic lights) and/or pedestrians in a fast and efficient way. The inherent proximity of the communications makes D2D transmissions (possibly exploiting multihop transmissions) one of the key enabling technologies for these services.

For this reason, we recently enhanced SimuLTE so as to make it interoperable with Veins [39,40], an OMNeT++-based framework for the simulation of vehicular networks, which is widely used by the research community. This way, it is possible to endow vehicles with an LTE NIC, immersing them into a cellular infrastructure where they can communicate with other network elements and/or among them, possibly exploiting D2D transmissions.

We consider the scenario depicted in Figure 26, which is taken from the 3GPP specifications [41]. The latter describes an urban scenario, defined as a grid of size 250 m × 433 m. Streets have two lanes per direction and each lane is 3.5 m wide. Inter-site distance between eNBs is 500 m. With reference to Figure 26, we defined four bidirectional vehicle flows, respectively connecting points A-B, C-D, E-F and G-H. From each entry point A to H, a new vehicle enters the network each 2.5 s. Vehicles are attached to best serving eNB according to a best-Reference-Signal-Received-Power (RSRP) criterion when they enter the grid, and they can perform handover to another eNB when they perceive a better RSRP. Vehicles moves at a speed of 60 Km/h. This means that the distance between vehicles in the same lane is 41.67 m. As shown in [42], these values for speed and distance correspond to a non-rush-hour scenario, although they are not very dissimilar to a rush-hour case, where speed and distance in the rush-hour are respectively 36 Km/h and 20 m. Physical layer parameters are the same as the previous section and presented in Table 1. We will analyze the broadcast delay with increasing target radius. For each configuration, we run a simulation for 250 s and, on each second, a randomly chosen vehicle generates one message and starts the broadcasting. Since vehicles appear in the network dynamically (hence, there are no vehicles in the network at the beginning of the simulation), we start gathering statistics after a warm-up period of 50 s.

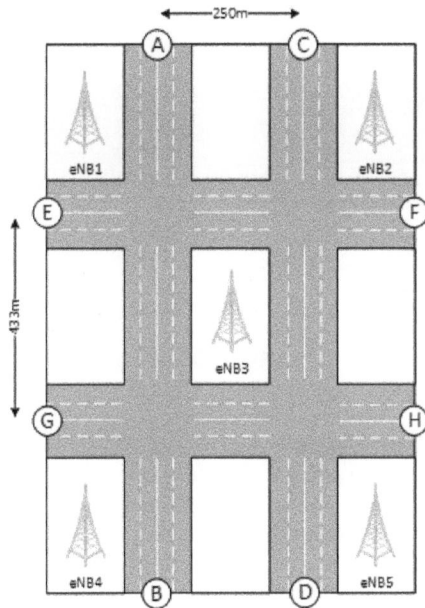

Figure 26. Urban grid scenario.

The Trickle suppression mechanism is enabled and vehicles relay a message if they are within the target area defined within the message. MaxC/I is employed as the scheduling algorithm, hence frequency reuse is not enabled. Resource allocation is performed according to the SRA mode, which proved to be more efficient, with vehicles transmitting using CQI = 7.

Figures 27 and 28 shows respectively the average and the 95th percentile of the delay required to complete the broadcasting, as a function of the target radius.

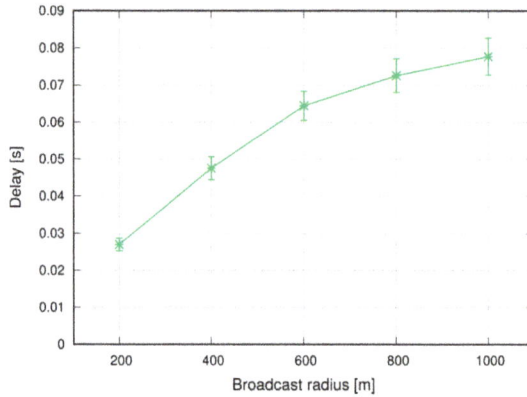

Figure 27. Average broadcasting delay.

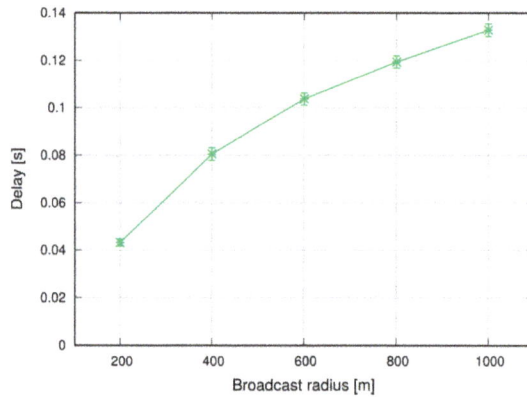

Figure 28. 95th percentile of broadcasting delay.

By comparing the results with those obtained in the previous section, we can observe that both average and 95th percentile of the delay are higher in the vehicular case, although not overly so. This is due to the particular deployment of the vehicles and the definition of the target area as a circle. With reference to Figure 29, we assume that the circled vehicle (bottom right of the figure) generates a message to be broadcast within the area defined by the dashed circle. In order to reach the vehicle highlighted by the square, the message needs to traverse several hops along the roads, since it is too far to be reached with a single, direct transmission and there are no other vehicles along the straight path between them.

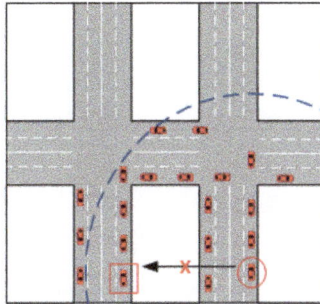

Figure 29. Example of broadcasting. The dashed circle delimits the broadcast area.

In any case, Figure 30 shows that the reliability of the dissemination is above 99%. Note that in this case vehicles are moving, hence they may enter/leave the target area while a broadcast is ongoing (although this effect is negligible in practice, given the short time it takes for a broadcast to cover the area). To maintain consistency, we count in the delivery ratio only the vehicles that are within the target area at time t_0. Figure 31 shows that the average number of RBs allocated for each broadcast, which is around 280 for a target radius of 1000 m. In this case, too, the number of RBs used is of the same order of the number of UEs, which is quite low.

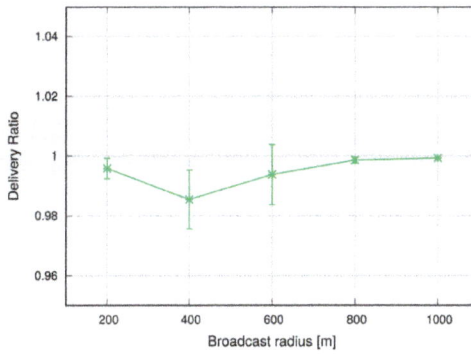

Figure 30. Average delivery ratio.

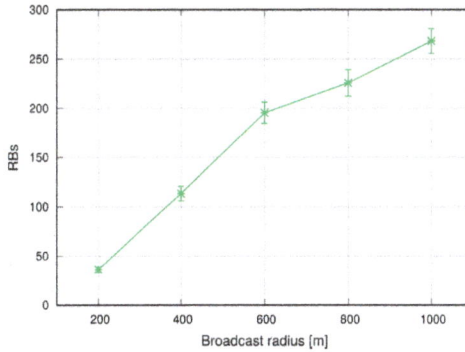

Figure 31. Allocated RBs per broadcast.

6. Conclusions

In this paper we have presented a solution for message broadcasting in LTE-A, using multihop P2MP D2D transmissions. This solution relies on application-level intelligence on the UEs, and leverages standard D2D resource allocation schemes in the network. It allows UEs to specify the target area, without being constrained by cell boundaries.

We performed simulations in both a static and a vehicular environment, in a multi-cell network. Our results show that this type of broadcast is fast, taking 80–120 ms to cover a 1000 m target radius at the 95th percentile. Moreover, it is highly reliable, meaning that the percentage of UEs actually reached by the message is close to 100%. Last, but not least, it is cheap in terms of resource consumption. In fact, it does not occupy the DL frame (the SL being carved out of the UL spectrum); this means that no service disruption occurs in the DL, and that no additional power is consumed by the eNB to support this service. Moreover, the amount of UL resources consumed is quite limited, thanks to the possibility of frequency reuse and proximate transmissions with higher CQIs. The RBs consumed for a broadcast, on average, are less than one per UE, which makes MDB quite economical, and minimally disruptive of other services that an LTE network would need to carry on simultaneously.

Further research on this topic, ongoing at the time of writing, includes at least two directions: first, investigating a deeper involvement of the infrastructure, in particular of the eNBs, in the broadcast relaying process. If the eNB is aware that a multihop transmission on the SL is being required by the sending applications, then it may allocate grants proactively to speed up the process. The second direction is to leverage network intelligence and network-wide information to characterize the target area. With reference to the vehicular case, the alert application could be made more efficient by being more selective as to which destination UEs it targets. For instance, if the message notifies that the originating vehicle is suddenly slamming on the brakes, the alert should reach the vehicles that are following it, and not those preceding it or across a block. This highlights the problem of building context-aware broadcast domains. On one hand, defining a more detailed area may occupy more space in the application-level message, which is something to consider carefully if the above-mentioned benefits are to be retained. On the other, we argue that a context-aware definition of the target area may not be defined by the vehicles themselves, since they may lack the knowledge of the surrounding environment and position of neighboring vehicles. Acquiring this knowledge using distributed means (i.e., inter vehicle communications) may not be viable either, because it would take a non-negligible message exchange and time, whereas broadcasting alerts should be accomplished in real-time to meet strict deadline requirements. The emerging MEC paradigm can play an important role in this respect. With MEC, vehicles could acquire the information about the intended geographical reach of one message by querying the corresponding service running at an application server located at the edge of the mobile network. The latter can leverage the location services provided by the network operator to define which vehicles should be receiving the message, and define a target area on behalf of the originator. Low latency would be guaranteed by the proximity of the MEC server to the vehicles, and by single client-server interaction.

Acknowledgments: The subject matter of this paper includes description of results of a joint research project carried out by Telecom Italia and the University of Pisa. Telecom Italia reserves all proprietary rights in any process, procedure, algorithm, article of manufacture, or other result of said project herein described.

Author Contributions: The three authors conceived and discussed the paper; Giovanni Nardini performed the experiments; the three authors analyzed the data; Giovanni Nardini and Giovanni Stea wrote the paper; Antonio Virdis revised and edited the paper.

Conflicts of Interest: The authors declare no conflict of interest. The funding sponsors had no role in the design of the study; in the collection, analyses, or interpretation of data; in the writing of the manuscript, and in the decision to publish the results.

References

1. Zanella, A.; Bui, N.; Castellani, A.; Vangelista, L.; Zorzi, M. Internet of Things for Smart Cities. *IEEE Int. Things J.* **2014**, *1*, 22–32. [CrossRef]
2. Seo, H.; Lee, K.-D.; Yasukawa, S.; Peng, Y.; Sartori, P. LTE Evolution for Vehicle-to-Everything (V2X) Services. *IEEE Commun. Mag.* **2016**, *54*, 22–28. [CrossRef]
3. Chmaj, G.; Selvaraj, H. Distributed processing applications for UAV/drones: A Survey. In *Progress in Systems Engineering*; Springer: Cham, Switzerland, 2015; Volume 366, pp. 449–454.
4. Bilstrup, K.; Uhlemann, E.; Strom, E.G.; Bilstrup, U. Evaluation of the IEEE 802.11p MAC Method for Vehicle-to-Vehicle Communication. In Proceedings of the VTC 2008-Fall IEEE 68th Vehicular Technology Conference, Calgary, BC, Canada, 21–24 September 2008; pp. 1–5. [CrossRef]
5. Araniti, G.; Campolo, C.; Condoluci, M.; Iera, A.; Molinaro, A. LTE for vehicular networking: A survey. *IEEE Commun. Mag.* **2015**, *51*, 148–157. [CrossRef]
6. Soleimani, H.; Boukerche, A. D2D scheme for vehicular safety applications in LTE advanced network. In Proceedings of the IEEE International Conference Communications (ICC), Paris, France, 21–25 May 2017.
7. 5G-PPP. 5G PPP Use Cases and Performance Evaluation Models. White Paper. Available online: https://5g-ppp.eu/wp-content/uploads/2014/02/5G-PPP-use-cases-and-performance-evaluation-modeling_v1.0.pdf (accessed on 1 November 2017).
8. ETSI GS MEC 013 v1.1.1. (2017-07): Mobile Edge Computing (MEC); Location API. Available online: http://www.etsi.org/deliver/etsi_gs/MEC/001_099/013/01.01.01_60/gs_MEC013v010101p.pdf (accessed on 1 November 2017).
9. Yu, C.H.; Tirkkonen, O.; Doppler, K.; Ribeiro, C. On the Performance of Device-to-Device Underlay Communication with Simple Power Control. In Proceedings of the VTC Spring 2009—IEEE 69th Vehicular Technology Conference, Barcelona, Spain, 26–29 April 2009.
10. Nardini, G.; Stea, G.; Virdis, A.; Sabella, D.; Caretti, M. Broadcasting in LTE-Advanced networks using multihop D2D communications. In Proceedings of the 2016 IEEE 27th Annual International Symposium on Personal, Indoor, and Mobile Radio Communications (PIMRC), Valencia, Spain, 4–8 September 2016. [CrossRef]
11. GPP—TR 36.843 v12.0.1. Study on LTE Device-to-Device Proximity Services: Radio Aspects (Release 12). 2014. Available online: http://www.3gpp.org/ftp/Specs/archive/36_series/36.843/36843-c01.zip (accessed on 1 November 2017).
12. Da Silva, J.M.B.; Fodor, G.; Maciel, T.F. Performance Analysis of Network-Assisted Two-Hop D2D Communications. In Proceedings of the Globecom Workshops (GC Wkshps), Austin, TX, USA, 8–12 December 2014.
13. Wang, S.Y.; Guo, W.S.; Zhou, Z.Y.; Wu, Y.; Chu, X.L. Outage Probability for Multi-Hop D2D Communications With Shortest Path Routing. *IEEE Commun. Lett.* **2015**, *19*, 1997–2000. [CrossRef]
14. Rigazzi, G.; Chiti, F.; Fantacci, R. Car Multi-hop D2D networking and resource management scheme for M2M communications over LTE-A systems. In Proceedings of the 2014 International Wireless Communications and Mobile Computing Conference (IWCMC), Nicosia, Cyprus, 4–8 August 2014.
15. Militano, L.; Orsino, A.; Araniti, G.; Molinaro, A.; Iera, A. A Constrained Coalition Formation Game for Multihop D2D Content Uploading. *IEEE Trans. Wirel. Commun.* **2016**, *15*, 2012–2024. [CrossRef]
16. Militano, L.; Condoluci, M.; Araniti, G.; Molinaro, A.; Iera, A.; Muntean, G.M. Single Frequency-Based Device-to-Device-Enhanced Video Delivery for Evolved Multimedia Broadcast and Multicast Services. *IEEE Trans. Broadcast.* **2015**, *61*, 263–278. [CrossRef]
17. Xia, Z.; Yan, J.; Liu, Y. Energy efficiency in multicast multihop D2D networks. In Proceedings of the 2016 IEEE/CIC International Conference on Communications in China (ICCC), Chengdu, China, 27–29 July 2016.
18. Lin, X.; Andrews, J.; Ghosh, A.; Ratasuk, R. An overview of 3GPP device-to-device proximity services. *IEEE Commun. Mag.* **2014**, *52*, 40–48. [CrossRef]
19. Cappanera, P.; Lenzini, L.; Lori, A.; Stea, G.; Vaglini, G. Optimal joint routing and link scheduling for real-time traffic in TDMA Wireless Mesh Networks. *Comput. Netw.* **2013**, *57*, 2301–2312. [CrossRef]
20. Draves, R.; Padhye, J.; Zill, B. Routing in multi-radio, multihop wireless mesh networks. In Proceedings of the Mobicom '04—10th Annual International Conference on Mobile Computing and Networking, Philadelphia, PA, USA, 36 September–1 October 2004; ACM: New York, NY, USA, 2004.

21. Williams, B.; Camp, T. Comparison of broadcasting techniques for mobile ad hoc networks. In Proceedings of the MOBIHOC '02—3rd ACM International Symposium on Mobile Ad Hoc Networking & Computing, Lausanne, Switzerland, 9–11 June 2002.

22. Kyasanur, P.; Choudhury, R.R.; Gupta, I. Smart Gossip: An Adaptive Gossip-based Broadcasting Service for Sensor Networks. In Proceedings of the 2006 IEEE International Conference on Mobile Adhoc and Sensor Systems (MASS), Vancouver, BC, Canada, 9–12 October 2006; pp. 91–100.

23. Mkwawa, I.H.M.; Kouvatsos, D.D. (2011) Broadcasting Methods in MANETS: An Overview. In *Network Performance Engineering. Lecture Notes in Computer Science*; Kouvatsos, D.D., Ed.; Springer: Berlin, Germany, 2011; Volume 5233.

24. Ruiz, P.; Bouvry, P. Survey on Broadcast Algorithms for Mobile Ad Hoc Networks. *ACM Comput. Surv.* **2015**, *48*, 8. [CrossRef]

25. Tseng, Y.-C.; Ni, S.-Y.; Shih, E.-Y. Adaptive approaches to relieving broadcast storms in a wireless multihop mobile ad hoc network. *IEEE Trans. Comput.* **2003**, *52*, 545–557. [CrossRef]

26. Leentvaar, K.; Flint, J. The Capture Effect in FM Receivers. *IEEE Trans. Commun.* **1976**, *24*, 531–539. [CrossRef]

27. Haus, M.; Waqas, M.; Ding, A.Y.; Li, Y.; Tarkoma, S.; Ott, J. Security and Privacy in Device-to-Device (D2D) Communication: A Review. *IEEE Commun. Surv. Tutor.* **2017**, *19*, 1054–1079. [CrossRef]

28. Chatzopoulos, D.; Ahmadi, M.; Kosta, S.; Hui, P. Have you asked your neighbors? A Hidden Market approach for device-to-device offloading. In Proceedings of the 2016 IEEE 17th International Symposium on a World of Wireless, Mobile and Multimedia Networks (WoWMoM), Coimbra, Portugal, 21–24 June 2016.

29. Chatzopoulos, D.; Ahmadi, M.; Kosta, S.; Hui, P. OPENRP: A reputation middleware for opportunistic crowd computing. *IEEE Commun. Mag.* **2016**, *54*, 115–121. [CrossRef]

30. Smith, M.J.D.; Goodchild, M.F.; Longley, P. *Geospatial Analysis: A Comprehensive Guide to Principles, Techniques and Software Tools*; Matador: Leicester, UK, 2007.

31. Levis, P.; Patel, N.; Culler, D.; Shenker, S. Trickle: A self-regulating algorithm for code propagation and maintenance in wireless sensor networks. In Proceedings of the 1st USENIX/ACM Symposium NSDI 2004, San Francisco, CA, USA, 29–31 March 2004; pp. 15–28. Available online: http://citeseerx.ist.psu.edu/viewdoc/download?doi=10.1.1.500.348&rep=rep1&type=pdf (accessed on 1 November 2017).

32. Nardini, G.; Stea, G.; Virdis, A.; Sabella, D.; Caretti, M. Resource allocation for network-controlled device-to-device communications in LTE-Advanced. *Wirel. Netw.* **2017**, *23*, 787–804. [CrossRef]

33. Cimmino, A.; Pecorella, T.; Fantacci, R.; Granelli, F.; Rahman, T.F.; Sacchi, C.; Carlini, C.; Harsh, P. The Role of Small Cell Technology in Future Smart City Applications. *Trans. Emerg. Telecommun. Technol.* **2014**, *25*, 11–20. [CrossRef]

34. Virdis, A.; Stea, G.; Nardini, G. Simulating LTE/LTE-Advanced Networks with SimuLTE. In *Advances in Intelligent Systems and Computing*; Springer Nature: London, UK, 2016; Volume 402, pp. 83–105. [CrossRef]

35. Varga, A.; Hornig, R. An overview of the OMNeT++ simulation environment. In Proceedings of the SIMUTools '08—1st International Conference on Simulation Tools and Techniques for Communications, Networks and Systems & Workshops, Marseille, France, 3–7 March 2008.

36. Nardini, G.; Virdis, A.; Stea, G. Simulating device-to-device communications in OMNeT++ with SimuLTE: Scenarios and configurations. In Proceedings of the 3rd OMNeT++ Community Summit, Brno, Czech, 15–16 September 2016.

37. 3GPP TR 36.814 v9.0.0. Further Advancements for E-UTRA Physical Layer Aspects (Release 9). 2010. Available online: http://www.3gpp.org/ftp/Specs/archive/36_series/36.814/36814-900.zip (accessed on 1 November 2017).

38. Zhou, X.; Zhang, Z.; Wang, G.; Yu, X.; Zhao, B.Y.; Zheng, H. Practical Conflict Graphs in the Wild. *IEEE/ACM Trans. Netw.* **2015**, *23*, 824–835. [CrossRef]

39. Sommer, C.; German, R.; Dressler, F. Bidirectionally Coupled Network and Road Traffic Simulation for Improved IVC Analysis. *IEEE Trans. Mob. Comput.* **2011**, *10*, 3–15. [CrossRef]

40. Nardini, G.; Virdis, A.; Stea, G. Simulating Cellular Communications in Vehicular Networks: Making SimuLTE Interoperable with Veins. In Proceedings of the 4th OMNeT++ Community Summit, Bremen, Germany, 7–8 September 2017.

41. 3GPP TR 36.885 v14.0.0. Study on LTE-Based V2X Services (Release 14). 2016. Available online: http://www.3gpp.org/ftp/Specs/archive/36_series/36.885/36885-e00.zip (accessed on 1 November 2017).
42. Wisitpongphan, N.; Bai, F.; Mudalige, P.; Sadekar, V.; Tonguz, O. Routing in Sparse Vehicular Ad Hoc Wireless Networks. *IEEE J. Sel. Areas Commun.* **2007**, *25*, 1538–1556. [CrossRef]

MDPI AG

St. Alban-Anlage 66

4052 Basel, Switzerland

Tel. +41 61 683 77 34

Fax +41 61 302 89 18

http://www.mdpi.com

FutureInternet Editorial Office

E-mail: futureinternet@mdpi.com

http://www.mdpi.com/journal/futureinternet

www.ingramcontent.com/pod-product-compliance
Lightning Source LLC
Chambersburg PA
CBHW051857210326
41597CB00033B/5928